Meditations
with Cows

Meditations with Cows

WHAT I'VE LEARNED FROM DAISY,
THE DAIRY COW WHO CHANGED MY LIFE

Shreve Stockton

A TarcherPerigee Book

tarcher perigee

An imprint of Penguin Random House LLC
penguinrandomhouse.com

First trade paperback edition 2022

Photographs courtesy of the author.

TarcherPerigee with tp colophon is a registered trademark of Penguin Random House LLC.

Most TarcherPerigee books are available at special quantity discounts
for bulk purchase for sales promotions, premiums, fund-raising, and educational needs.
Special books or book excerpts also can be created to fit specific needs.
For details, write: SpecialMarkets@penguinrandomhouse.com.

ISBN 9780593086674 (hardcover)
ISBN 9780593086698 (ebook)
ISBN 9780593086681 (paperback)

Printed in Mexico
1   3   5   7   9   10   8   6   4   2

Book design by Laura K. Corless

for my mothers

Meditations
with Cows

Wisdom begins in wonder.

—SOCRATES

CHAPTER 1

Becoming Family

Back in high school, I was voted "Most Likely to Wake Up in a Strange Place." I've had an adventurous life and awoken in many a strange place: tucked into seats of Greyhounds going anywhere during my teenage years; on a boat in the Santa Barbara harbor where I lived illegally my first two years of college; alone in desolate Death Valley in the heat of early summer, reckoning with my mortality during a health crisis in my twenties. But perhaps the strangest place of all is on the warm, broad back of a two-thousand-pound Black Angus bull. I woke up on his back because I had fallen asleep there, stretched out against his heat, my body gently rising and falling on the tide of his breath as he chewed his cud in the winter sun.

I didn't grow up with cattle. I'd never been face to face with a bovine until I was twenty-eight years old, newly moved to Wyoming after becoming enchanted by the Bighorn Mountains during a solo ride from San Francisco to New York City on my Vespa. Instead of moving back to New York as I'd planned, I stayed a few weeks with a friend in Brooklyn, found a Wyoming rental on Craigslist, and moved, sight unseen, to a tiny town I'd never heard of.

I got a job as a substitute teacher in the local school—not quite a one-room schoolhouse, but not far from it, with less than ten kids in each grade. After spending my days under the fluorescent lights of the classroom, I spent afternoons afoot, exploring my new surroundings. One day, on the cusp of spring, I was jogging down the dirt road north of my rental house when I spotted a baby calf nestled under a sagebrush. Since moving to Wyoming, I'd seen herds of cattle grazing great pastures but had never seen a calf up close. I peered at the calf from across the fence that bordered the road, not sure if it was sleeping or sick or dead; not knowing that mother cows often hide their newborns under bushes and sagebrush, and that cattle curl up to nap in a pose similar to that of a resting dog.

A dusty pickup truck pulled up beside me. The driver rolled down his window. The two cowboys in the cab introduced themselves as Mark and Mike, both clearly amused to find someone staring so intently at a commonplace calf. "That's a baby Black Angus," Mark said. "It's Mike's—wanna hop in and help us feed? You can meet the big ones." The bed of the truck was piled high with hay bales. My curiosity piqued, I squeezed in with them. We drove through a gate and out into the middle of a pasture, then Mark idled the truck while Mike and I got out. Mike climbed onto the stack of bales and hoisted me up next to him. As Mark drove slowly across the pasture, Mike showed me how to cut the baling twine and kick off flakes of hay to the swarm of black cows that had appeared behind the truck. I helped Mike feed his cows the next day, and the day after that. By the following week, we had a standing hay date.

Our budding friendship bloomed into more, and when my rental lease was up in October, I moved into a little log cabin at the edge of Mike's forty-acre property with Eli, the feral cat I had adopted. The cabin was a single room measuring twelve feet by twelve feet, which Mike had built using little more than a chainsaw and hammer. Originally meant as a tack shed, the cabin had rough plywood floors and no real kitchen, no sink—just a small countertop in one corner to which I added a mini fridge, a hotplate, and a toaster oven. A lean-to on the east side of the cabin sheltered a shower and toilet, functional in all but the coldest weather. I got drinking water from the hose that fed a cow trough nearby and washed my dishes in the shower.

The only heat in the cabin came from an old cast-iron woodstove. I learned how to chop wood and, over time, learned the character of different woods. I burned cedar on days when I worked outside so I could enjoy the deliciously scented smoke wafting from the stovepipe, perfuming the air. I banked the fire with cottonwood when I went to town—though it was abundant, I did not enjoy the smell of cottonwood. I saved the largest logs to use at night; since they burned the longest, I didn't have to wake every few hours to feed the fire. I had a special stack of heavy blocks of pine filled with crystalline pitch, reserved for the coldest mornings. The pitch gleamed like gold and was just as precious, for it was extremely flammable, and those logs burst into flame immediately and burned extra hot.

Sometimes, my fire went out in the night and I'd wake to a freezing cabin with Eli curled up under the covers beside me. I learned to leave my teakettle half full of water in the evening and if the cabin froze, I clunked the ice-filled teapot onto the woodstove to thaw over the morning's fire. As the ice inside melted, I poured out a little water to brush my teeth. When the water warmed up, I poured a little more on a washcloth to wash my face. Then it would boil, and I'd make coffee. Mike thought I was crazy to choose to live in the cabin and tried to talk me into moving in with him instead. But I had fallen for that tiny cabin the first time I drove past it on the road, months before I even met Mike. It had called to me, and in answering, I found the only place I'd ever lived that truly felt like home.

The following spring, Mike stopped by the cabin one April afternoon, reached into the pocket of his coat, and carefully, solemnly, held out his hand. In his palm was an orphaned coyote pup. The baby coyote couldn't have been more than ten days old, his eyes barely open. With much initial trepidation, I took the wild orphan into my home and my heart. At the time, I believed it would be temporary—that I would care for the helpless pup, who we named Charlie, until he could care for himself in the wild. But rehabbing animals for release is strategic work— work I knew nothing about—and it does not entail allowing a coyote to sleep in one's bed or fraternize with one's cat. As the months went by, I realized it would be grossly irresponsible to release Charlie into the wild—to do so would have been unsafe for those in my community and disastrous for Charlie. From the very beginning, Charlie and I were copilots, inextricably each other's, bound by trust and duty and love.

During our first year together, Charlie transformed me in a way I never thought possible: he grounded me. Charlie tamed me as much as I tamed him, and by the time his birthday rolled around, I had, for the first time in my life, set down tentative yet tenacious roots. I no longer wanted to roam. Ironically, relinquishing my freedom to prioritize Charlie opened a new realm of possibility—the opportunity to fulfill a fantasy I'd held for as long as I could remember. With my heart tethered to Charlie and Wyoming, I could commit to a dairy cow.

I dreamed of having a dairy cow when I was a teenager, incongruous as it was to my wanderlust. I dreamed of having a dairy cow when I lived in the city, impractical as that would have been. When I read a *New Yorker* profile on a cow-milking, cheese-making nun while sitting on my roof in San Francisco, part of me longed to be her (minus the piousness). And while this longing began as a fanciful wish, it was also rooted in the practical. I found raw milk more nourishing and easier to digest than conventional pasteurized milk, preferred it enough to find a black market source for raw milk as soon as I moved to Wyoming. At the time, buying and selling raw milk, fresh from a cow or goat, was illegal in the state, as it still is in many others.

But first I had to find a dairy cow. The ranches in my county were home to beef cattle, and my illicit milk came from goats. I asked my veterinarian if he knew of anyone with a dairy cow for sale. He put me in touch with Lynn, the owner of a small commercial dairy a hundred miles to the north, who was selling several cows that did not produce as much milk as the rest of his herd. A "low producer" in the eyes of a dairy was perfect for me, so Mike and I drove out with the horse trailer in tow. Lynn gave us a full tour of the dairy—the milking parlor lined with stanchions and milking machines, the cooling and storage facility that housed enormous stainless-steel holding tanks of milk. Then he led us outside to meet the cows.

We were immediately approached by a huge white cow with dark brown eyes and a tuft of blond hair on the top of her head. As she strode up to us, Lynn introduced her as Daisy. Though the rest of the cows at the dairy were known by numbers, Daisy was so charming they had named her. "We're awful fond of her," Lynn said as he casually rubbed her back, "but she only averages five gallons of milk a day, which is half of what our other cows produce, and we just can't afford to keep her." Daisy reached out to sniff my neck, then looked down her nose at me, sunlight glinting off her long, white eyelashes. I offered her my hand to smell and she ignored it.

Daisy was three years old, three-quarters Brown Swiss and one-quarter Jersey. She'd had her second calf a few months earlier, so she was producing milk, and she was pregnant again, due in December. Like all mammals, dairy cows make milk when they have babies. While beef cows in Wyoming are bred to calve (give birth to their calves) in the spring, dairies stagger their calving throughout the year so they have a consistent supply of milk to sell year-round. Lynn led us farther into the corrals to show us half a dozen other cows, all statuesque black-and-white Holsteins. Daisy swaggered along confidently beside us while Lynn pointed out cows and rattled off the statistics of milk production and age.

After making a large loop through the corrals, we returned to the main gate, Daisy still beside us. Lynn and Mike looked at me expectantly, awaiting my verdict. Did I want one of these cows, or did I want to keep looking? And if I did want one of them, which one? Daisy held me in her gaze as my mind raced through the information I'd just heard. I had never met a dairy cow prior to this excursion. I'd done some research online, but I had no experience from which to make this decision. So I made the decision the way I make my best decisions. I went with my intuition. I went with Daisy.

We went to Lynn's office to do the paperwork, then everyone at the dairy came out to say farewell to Daisy before we loaded her in Mike's horse trailer. They were genuinely sad to see her go. We drove home slowly, past the patchwork of sugar beet and barley fields, along the sparkling river lined with trees unfurling in blossoms and leaves, to the rust red buttes of home. We pulled up to Mike's corral in early afternoon. I wanted to give Daisy a few days in the corral to get used to me and adjust to her new home before letting her roam Mike's property. Daisy sauntered from the trailer and began exploring the corral, sniffing out her new surroundings. "You have a cow!" Mike said, hugging me.

"I have a cow," I smiled back. "Thanks for going with me. I'm glad you were there." When Mike left to park the horse trailer, I filled the water trough for Daisy and brought her an armload of hay. I climbed into the feed bunk and sat cross-legged beside the pile of hay while Daisy ate, talking to her, telling her about her new home, then drifting into silence, listening to her chew and sigh. As she ate, I studied her body, her demeanor. She was relaxed and calm. Her coat was smooth and soft. Peach skin rimmed her eyes. She had been branded on her left side, and the scar looked like the Big Dipper. She had a blue ear tag in each ear, stiff flags the size of my palm displaying the number she'd been assigned before receiving a name. I cut the buttons off the back of the tags and pulled them out. She didn't shy away from me as I fiddled with her ears, and when I finished, she rubbed her forehead rhythmically against my torso and nearly pushed me out of the bunk. When I reached out to pet her face, Daisy maneuvered her head under my hand, guiding me to scratch above her left eye. She smelled like warm pound cake, like laundry dried in the sun, surprisingly sweet, more floral than the scent of horses, more delicate than that of dogs.

The sun drifted across the western sky as we lounged in the corral together. Dusk was still a few hours away, but I knew I needed to start milking. Since Daisy was making milk, she had to be milked morning and evening, effective immediately. I had never milked an animal before, and Daisy had never been milked by hand. I felt the butterflies of the unknown stirring in my belly. My companions across decades and continents, their flutters still made my breath catch in my throat. Each minute I stalled was a minute of daylight lost. "You ready to try this, Daisy?" I asked. I took her nonchalance as a yes and ran to the cabin to collect the stainless-steel milk pail I had preemptively bought one hopeful night online. I gathered up a few washcloths, a pail of warm water, a curry comb, and a soft-bristled horse brush, and shuffled back down to the corral.

Walking backward with a flake of hay, I lured Daisy into a small pen within the corral. I set the pails and brushes and rags on the ground beside her. I brushed her belly and sides with the curry comb as an offering of devotion and to release any loose hair. Daisy sighed as I brushed her, half closed her eyes. I used the soft brush on her pendulous udder, which was warm and dense and covered in a fine coat of white hair. I crouched in the dirt beside Daisy's hind leg, squatting on my heels so I could spring backward if she kicked. I washed her teats with a warm washcloth, patting them dry with another. Her teats were a ruddy peach, nearly the same color as my hands. I grasped one gently, my mind blank but for two prayers on repeat: *Please don't let me hurt her . . . Please don't let her kick me.* Daisy shifted her weight when I started squeezing her teat, but she didn't balk. She remained focused on the hay I'd piled in front of her.

The mechanics of extracting milk turned out to be surprisingly easy to figure out. It was not a straight squeeze; more like the movement of rapidly rapping your fingers on a desk, yet doing so while your hand is in a fist. But the act of milking was considerably slower than I'd anticipated. My hands tired quickly, and I milked with one hand at a time so the alternate hand could rest. Daisy was patient with me. She expressed minor discomforts by shifting her weight. I soon realized that when she had her weight on the hind leg by which I sat, it was impossible for her to kick me without first shifting her weight to

11

the other leg, and understanding this helped me begin to relax. But I was overly paranoid. Though she swatted her tail in annoyance when I accidentally pulled her udder hair when grasping a fresh teat, she was calm throughout my fumbling and never did kick.

I sipped some of her fresh milk as I worked. Sweet and frothy, it was the same temperature as the interior of my mouth. It comforted me, calmed the panic that began to rise as the sun began to set. I was sitting in the dirt, two teats into a four teat job, and it was about to get dark. I wondered if perhaps I was in over my head. A few slow sips of Daisy's milk reminded me that I was sitting in the dirt beside a beautiful white cow and she was letting me learn how to milk.

Even milking as quickly as my burning forearms would allow, the sun went down before I finished. I swore I could hear Daisy producing milk faster than I could milk it out. When twilight settled into night and it got too dark to see what I was doing, I turned on the lights at the corral and continued to milk. Cramped and cold and overwhelmed, I eventually squeezed the last drops of milk from her teats. I let Daisy out of the little pen and brushed her with the curry comb, from her face to her flanks and down to her ankles. As I brushed every gleaming inch of her body, I told her how grateful I was for her patience with me, that I was so glad she was here with us, that I hoped she'd sleep well, and that I'd see her in the morning.

I walked back to the cabin, lugging the heavy milk pail and the empty water pail and the brushes and washcloths. I got a quick fire going in the woodstove, then strained Daisy's milk through a mesh coffee filter into canning jars of various sizes. Nearly three gallons of milk, like bottled moonlight, covered my tiny countertop. It didn't all fit in my mini fridge, so I guzzled a quart for dinner and poured some out for Charlie, Eli, and Chloe, the border collie/hound dog I had adopted when Charlie was a year old. They drank in focused delight, a chorus of staccato laps. I called Mike to say good night and give him a brief recap of my milking misadventure, then collapsed into bed.

Milking is both ritual and chore, which extends beyond the actual act of milking. There is the washing of the milk pail and filter and glass storage jars every day. There is the cow wrangling and the udder washing and the post-milking thank-you brushing of the cow. There is the toting of the milk pail and cup and brushes and washcloths and warm soapy water and hay and treats to the milking area, and the lugging of it all, plus twenty pounds of milk, back home again when done (but for the water, hay, and treats). And there is the task of making sure the cow is peaceful and happy or distracted and enter-tained while she is being milked and, even if she is none of those things, that she at least stands still.

I found myself in awe of the intimacy of milking, of the rela-tionship that Daisy and I forged through this daily routine, of the connection inherent in the act. When you have a milk cow, you are together every day, no matter the weather, no matter either of your moods. The hind leg of this twelve-hundred-pound animal towers over you as you crouch beside her, her hoof solid as stone, both of you aware of the fact that one well-aimed kick could kill you if she wished. Yet you are allowed to rest your cheek and forehead against her warm belly as you milk. Your breathing syncs with hers. Her teats in your hands, her milk in your fridge, her trust in you entwined with your trust in her, you become family.

Imagine filling a five-gallon bucket with a ninety-nine-cent water gun. This approximates the time and hand strain involved in those early days of milking Daisy. I never realized how many individual muscles make up the human hand until I started milking Daisy, and every single one of those muscles burned and ached. The pain was excruciating—not just while I was milking, but constantly, unrelentingly. Yet I couldn't take a day off from milking, or even an afternoon, to let the muscles in my hands rest and repair themselves. As the gallons of milk filled my fridge and then Mike's, multiplying by the day, I popped ibuprofen and arnica, soaked my hands in bowls of hot water and Epsom salt, slathered them with warming balms, all to no avail. Desperate for help, I asked Mike if he knew of anyone with an orphan calf for sale.

Cows usually give birth to a single calf, but sometimes, a cow will have twins. When a cow delivers twins, she will almost always care for just one of the calves and abandon the twin. This isn't because cows are bad mothers—it's quite the opposite. A beef cow instinctually knows she only produces enough milk to raise one healthy calf, and she will not allow a second calf to nurse. Orphaned twins are bottle-fed or given to another cow who may have lost her calf to stillbirth, illness, or bad weather. If a rancher doesn't have a cow who can adopt the orphan calf, and doesn't want to bottle-feed it, they will sell the calf to someone who does. If I could find a calf for Daisy to adopt, that calf could help take some of the pressure off my aching hands by consuming a portion of Daisy's milk. Mike made a few calls, and the next afternoon, we drove to a neighbor's house to pick up an orphan.

A few weeks prior, one of our neighbor's cows had twins. The cow took care of one calf and orphaned the other. In a highly unusual turn of events, another of his cows, who had already calved and nursed her baby for a week, adopted the twin and orphaned her own calf. The rancher had been bottle-feeding the orphan and was happy to sell him to me. We hoisted the calf into the backseat of Mike's truck where he stood, serenely incongruous among the coveralls and toolkits. Since he had been bottle-fed, the calf was socialized to humans, and he let me pet his face and scratch his back during the short drive home to Daisy.

I wasn't sure what to expect when I introduced Daisy and the calf, so I stood back and watched their interaction. The calf took one look at Daisy's glorious udder, ran to her side, and nuzzled up to suckle her. Daisy, flabbergasted by the tiny beast, kicked him away and walked to the other side of the corral. Though Daisy had birthed two calves, she had never nursed a calf before. At the dairy, calves were separated from their mothers immediately. The calves were fed their mothers' milk from a bottle, and the cows were milked exclusively by machine. Daisy's reaction was completely understandable. I wasn't about to give up on the pairing, but I didn't want to pressure Daisy. I milked her as usual with my aching hands, poured some of her warm milk into a calf bottle to feed the hungry baby, and left them together in the corral for the night.

I continued to milk Daisy twice a day and fed our little calf, who

I called Baby, with her milk from a bottle. Afterward, I brushed him. Daisy, who never tired of getting brushed, had to sidle up next to Baby in her attempts to get my attention away from him. When I gave her a turn under the brush, I kept a hand on Baby, scratching his back so he would stay close to Daisy while I groomed her. When I brought Daisy armfuls of hay, I cuddled with Baby beside the feed bunk so the two of them brushed against each other's bodies while Daisy ate. They had slept at opposite ends of the corral, but I noticed, as the days went by, that they laid closer and closer to one another. One morning, on my way to the corral to milk at first light, I caught sight of Baby tentatively approaching Daisy. I hid behind Mike's backhoe to watch. Baby sidled up to her flank and extended his serpentine tongue, wrapped it around a teat, and drew the teat into his mouth. Daisy swung her head around to look back at him and shifted her weight, one of her familiar signs of mild annoyance, but she didn't kick him away. Baby's tongue enveloped the teat and he latched on. Daisy swished her tail, but she did not kick. And then Baby drank heartily, single-mindedly, until frothy milk slobber dribbled from the sides of his mouth.

With that first suckle, Daisy's suppressed maternal instincts bloomed like a Wyoming sunrise: epic, encompassing, brilliant. She adopted Baby wholly and completely as her calf and stood perfectly still, with an expression of serene fulfillment, as Baby nestled against her side and drank whenever he desired, both their eyes drifting closed during the act, peaceful and blissed out. She communicated with him through the various moos I had come to recognize from Mike's cows: a gentle, melodic moo meant "Baby, I love you"; a low, murmuring moo meant

"Baby, where are you?"; a louder, sharper moo meant "Baby, come this way"; and an urgent, deafening *MOOOOO!!* meant "Kiddo, if you don't get over here by the time I count to five . . ." She bathed him with her tongue, and he followed her everywhere, thrilled to be hers.

After a couple of months of milking daily, my hands grew strong and accustomed to the work. Milking took twenty minutes? Half an hour? Forty minutes? More? I never timed it and couldn't begin to guess. I felt untethered from time while I was with Daisy. During our milking sessions, I stood periodically to rub her head or rub my face against her cheek. I milked early, just as the sun was rising, the air still cool from the night, Daisy standing calmly above me against the blue and golden dawn. Sometimes, the wind was relentless and blew dirt into my milk pail no matter how I tried to shelter it, and that milk went to Charlie, Chloe, Eli, and Mike's chickens. Sometimes, Daisy and I were not in sync. Sometimes I was tired; sometimes she was antsy. Sometimes I was distracted and she showed her irritation by swishing her tail and stomping her hooves. But most of the time, it was a mutual meditation. I paused at intervals to sip her warm milk from a cup, as I had our first evening. Sometimes, I brought a thermos of coffee to the barn with me, and scooped the rich froth from the morning's milk into my coffee for a farm-fresh latte, sipped while watching Daisy licking Baby, the birds in flight, the day beginning.

Daisy's milk was a revelation, ecstasy in liquid form, with cream so thick I had to skim it with a spoon. With the nearest grocery store thirty miles away, restaurants rare and rarely open, and delivery non-existent, on particularly busy days, my dinner was nothing but a quart of Daisy's fresh milk, and I didn't mind at all—those meals of milk were luxurious in their simplicity, pure and rich and delicious. Even with Baby helping me, I milked more than Mike and I could consume. Jars of milk and cream filled my little fridge to the exclusion of anything else, and the overflow piled up in Mike's full-size fridge. We couldn't drink it as swiftly as I was milking it. Two gallons a day, every single day, adds up quickly. I didn't have a grasp on compound interest until the principle was illustrated with compounding milk.

I made yogurt by the gallon and tried my hand at cheese. I made butter by pouring cream into a mason jar, capping it tightly, and shaking the jar violently until my arms got sore, until I started to wonder if the method was madness or myth, until the cream bubbled and frothed and thickened into whipped cream. I kept shaking, until the thick mixture slackened again, thinned out, seemed to be going backward. And then, midshake, I felt the presence of butter before I saw it. The contents of the jar had a different kind of density. Through the glass, tiny golden orbs floated in a matrix of watery milk. I shook the jar until the orbs joined together and formed a large golden glob. Butter. I strained the buttermilk into a glass—unlike buttermilk from the store, this was light and sweet—and plopped the butter blob into a bowl. The color of a dandelion flower, it gleamed like sunlight, as if it were glowing from within, more luminous than any jewel. Butter is perhaps the most magical manifestation of sunlight: sunlight, transformed into grass through photosynthesis; that grass nourishing my cow; my cow metabolizing that energy into milk and sharing it with me; the rich, nutrient-dense fat of that milk separating into cream; that cream, when confronted with the force of motion, emulsifying into gold, edible gold, velvet on the tongue, sweet and soft and satisfying.

And still the milk piled up, gallon jars crammed together, bowing the shelves of Mike's fridge. I never considered selling the extra milk—legally and logistically, it was too overwhelming—but I gave it away to those who asked. Even after giving it away, and feeding it to every ecstatic cat and chicken and canine on the place, and making yogurt and cheese and butter, and Mike and I drinking it with every meal, I ran out of room to store it. So one evening, I drew myself a bath in the cast-iron tub I had in my garden, poured three gallons of milk into the water, and bathed in it. Cleopatra knew what she was talking about.

If Daisy felt an adjustment period, she didn't show it. She followed me closely—so closely, that to get photographs of her, I had to sprint ahead, spin around, and snap a few pictures in the seconds before she ambled up, angling in for a body rub. I could lead her anywhere with the brush in my hand. When I arrived home from errands and saw her grazing in the pasture by the road, I'd roll down my window and yell, "Hi, Daisyyyyy!" and she'd lift her head and give me a nod. She came when I called her, and we developed our own secret language of hand signals. Daisy was as smart and as sure as any horse or dog. Cattle are commonly thought of as dumb beasts, but that is so far from the truth. I wonder if the disconnect persists because it makes it easier for people to ignore how cattle are so often treated. Cattle are intelligent, and they are particularly brilliant at noticing and understanding patterns. They learn quickly and have phenomenal memories. Show a bovine—of any age—a routine or pattern just a few times, and they will commit it to memory and never forget.

After Daisy allowed Baby to nurse, I let them out of the corral to roam and graze all day, then separated them at night so Baby wouldn't drain Daisy's udder before I had a chance to milk in the morning. In the evenings, I gathered Daisy and Baby from the pasture and walked them down to the corral. Daisy followed me and Baby followed Daisy. I kept Baby in the corral overnight, and Daisy was free to spend the night in the pasture beside the corral, separated from Baby only by a rail fence, or roam as she wished. When I milked in the morning, I always left a teat untouched for Baby, and when I opened the gate to let him out of the corral, he galloped to Daisy's side, guzzled his breakfast, then nursed as he wished throughout the day.

One evening, I was so focused on work I didn't notice the time. A demanding *MOOOOOOOO!* shattered my concentration—a moo so thunderous, I felt the vibrations ripple through my body and bones. I opened the door to find Daisy staring at me, a foot from the door. She had walked in from the far reaches of the field to remind me that we had a schedule to keep. Behind her, Baby frolicked in the driveway. The sun floated just above the horizon. It was later than I'd realized.

"Daisy, is it time to tuck you in?" I asked, as I rubbed her forehead

and headed out toward the corral. Daisy followed me, her head bob-
bing at my shoulder, in step with my pace. Baby trailed after us. We
walked down the draw and up the other side, took the narrow trail
around the curve of the hill, and crossed to the back pasture. When
we reached the gentle slope that led to the corral, Daisy broke into a
trot, her udder swinging side to side. Baby ran after her, galloping a
wide arc into the pasture to build up speed, and raced through the
open gate into the corral, bucking in delight. As I brushed Baby, a
pheasant scurried through the corral, looking for bugs. I kissed Baby's
forehead, closed the corral gate behind me, and led Daisy out into the
adjoining pasture where I brushed her and scratched her favorite spot
above her left eye. "Good night, Daisy. Good night, Baby," I called to
them as I headed back to the cabin. "I love you."

One night, I woke at three in the morning with my mind racing.
Tentacles of tension wound around my trachea, my viscera. Stress
had crept into my shoulders and jaw, and anxiety made a home in my
chest, invading the space my breath was meant to fill. I tried to narrow
down the cause of my angst, find the reason behind it to then find the
cure, but it was nothing in particular and everything at once. I got out
of bed and dressed, knowing sleep was futile. I left Charlie and Chloe
curled on their mattress and slipped out into the darkness of deep
night. The silence and the darkness were equally vast; the silence
broken only by the occasional *whooovvo* of an owl perched somewhere

above; the darkness fractured only by the moon and the canopy of stars. The quarter moon reflected enough light to walk by, as long as I walked slowly, and I ventured out into the pasture behind my cabin. Daisy lay in repose at the far end of the field, the arc of her body glowing like the moon in the black expanse of pasture. As I approached her with careful steps, I heard her steady breathing, the silence parting briefly with each exhale. From the position of her head and ears, it was clear her eyes were on me, though I couldn't see such detail from a distance in the dark. As I got closer to her, I felt my stress wafting away behind me, dissipating like smoke, like layers of ghosts. The rhythm of her breaths guided and steadied my own. When I reached her side, I pressed my palms to her shoulder and stroked her soft neck. As I touched her, my breathing deepened further. My anxiety released with every exhale, as if it were being drawn out of me, absorbed by her body and grounded down into the earth. I rubbed Daisy's forehead and down her neck and spine, down her great, graceful back. She remained lying down and began chewing her cud in rhythmic nonchalance. In the power of Daisy's presence, the tension in my chest uncoiled, replaced by the buoyancy of my breath. In her presence, I was wholly present—my mind stopped wandering toward worry or fear or anger or blame or shame, those cryptic voids where we can spend so much time. With Daisy, I was simply me, the living being beyond the noise of my brain, with nothing to do or to prove. In the womb of night, Daisy guided this meditation as only the best gurus can—with her breath, with her presence, by her silent example. I bent to kiss Daisy's cheek in the moonlight, inhaled her warm scent, pressed my face against her neck as the stars sparkled above us. She reclined before me like a deity, her eyes softly closed, her eyelashes cradling moonbeams. The barnacles of negativity within me had transformed to gentleness, like a cascade of flower petals. I couldn't do this on my own. I couldn't let go like this on my own. Daisy's presence untangled me, pulled the dirty, sticky cobwebs of disappointment and fury from my shadows, dangled them in front of my face so I could see them, limp and dusty, and with a deep breath, blow them away into the silence surrounding us. My body was mine again, my mind quiet and clear. I leaned back against Daisy's resting body, slid down the slope of her shoulder, and nestled on the ground beside her, cuddling into the crook of her neck beneath the stars.

CHAPTER 2

Harmony and Hope

B aby gained over a hundred pounds every month through the spring and summer on his share of Daisy's decadent milk. In late September, when the light softened with the changing seasons and the cottonwood leaves began to dry and drop, I stopped milking Daisy and weaned Baby to give Daisy a rest period, known as "drying off," leading up to the birth of her new calf. By that time, Baby was half Daisy's size, stocky and strong, already developing the thick neck and shoulders of a bull. While all cattle are colloquially called cows, this terminology is technically incorrect. A cow is an adult female—and it makes poetic sense that we use the feminine descriptor as the default for a species with a matriarchal social structure. A male bovine who has been castrated is called a steer; those who have not are bulls. Females who have never had a calf—the maidens of the bovine world—are called heifers. Once they have their first calf, they become cows. Baby was an intact bull calf when we adopted him; he had not yet been castrated, the common fate for most male calves. Since he was adopted and his genetics were outside Mike's herd, we chose to let Baby grow up to be a bull. He would be old enough to breed cows the following year.

Autumn days turned into winter nights. Spiders crept through the cracks in the cabin walls and moved inside. I traded the daily milkshakes of summer for elk stew simmered on the woodstove, and drank my coffee black since I would be without cream until Daisy calved. By the end of November, snow had piled up in drifts, shrouding even the sagebrush. Daisy's coat had grown thick, her belly wide, and her udder ballooned to extraordinary proportions in preparation for the arrival of the calf she carried.

I had helped Mike with his cows enough to learn the visual clues that help to predict when a cow is getting close to calving. One of the most obvious signs is her udder. As she progresses through her pregnancy, a cow's udder will grow, and the skin of the udder gets increasingly taut. A vertical crease, or wrinkle, naturally bisects the back of a cow's udder when it's not completely full of milk. I'd learned from watching Mike's Angus cows that when the back wrinkle filled out and the udder appeared smooth and spherical, labor was imminent.

Daisy's udder was enormous—many times the size of an Angus

cow's—but the back crease in her udder was still prominent. Since I had never been a doula for a dairy cow, I wondered if perhaps the crease wouldn't fill out before she calved, or at all. Even with the wrinkle present, the volume of Daisy's udder far exceeded that of an Angus cow. Her graceful strides veered more toward a waddle with each passing day, in part because her calf in utero was getting larger, but mostly due to her bulging udder impeding her gait. When I got Daisy, Lynn had said she was due mid-December. But since her udder was already alarmingly large, and since babies don't always abide by the calendar, I moved Daisy into the little barn at the corral around the first of December and started watching her obsessively.

Theoretically, I didn't need to be present when Daisy calved. This would be her third calf. We always kept a close eye on heifers, for they may need assistance while calving or during the crucial stage when their calf tries to nurse for the first time, but experienced cows generally do perfectly well on their own, and many prefer not to have an audience. I cherished every opportunity I had to witness the profound act of birth, and with Mike's cows, I usually did so from a distance with binoculars, so as not to be intrusive to the mother. But Daisy was my first cow, having her first calf in my care. And though uncommon, complications during delivery can be fatal to the cow or calf. And it was December. Hoarfrost glittered on the bare branches of the cottonwood trees. When the sunset sky deepened to plum then darkened to black, the temperature dropped from below freezing to below 0°F. Calves are wet when they are born, and there was a real chance that Daisy's newborn could freeze to death, even in the barn in the bedding of fresh straw I had laid out. Being present while Daisy calved—bearing witness and being there to help should she need assistance—became my top priority.

When Charlie came into my life—two years to the day before Daisy joined us—I started a blog, posting stories and photos of Charlie online every day, which I have continued for the past thirteen years. Within months, my blog went viral, which led to a book deal, and to write my book, I quit my teaching job. Thanks to the book, which had been published six months before I got Daisy, I had savings in the bank for the first time in my adult life; and through the blog, my writing and photography were paying my bills. I knew how tenuous it was to be making a living from my art, and I worked more than I ever had with a day job. But in self-employment, I had gained flexibility in where and when I did my work, and this freedom allowed me to spend the majority of my time at the barn with Daisy.

When I was at the cabin, my thoughts were on Daisy. I woke in the night wondering what might be happening at the barn. And so I layered up in silk and wool, pulled on snow boots and gloves, and trekked down to the corral several times throughout each night to check on her. I lingered in the barn, waiting to see if Daisy licked her belly and sides, or held her tail out rigidly behind her, or stood up and lay down often and in succession—all signs of labor. But Daisy just lounged in the straw, peacefully chewing her cud. I brushed her fluffy winter coat, carried her armfuls of sun-dried grass hay that held the scent of summer, noted the ever-expanding growth of her udder, and returned to the cabin until curiosity demanded I visit Daisy once again.

The temperature plunged to -20°F at night, rising only to -14°F when the sun blazed down from a sapphire sky. I have a remarkable endurance for cold, though it didn't come naturally. In my youth, I cranked the heat when my parents weren't home and did my homework while lying directly on the heater vents. I developed my tolerance during my cross-country Vespa trip. The bliss of riding is that there are no barriers between you and the world through which you ride. My Vespa had no roof or windows—it didn't even have a windshield. Flying along, a foot above the ground, changes in air temperature wrapped around my body as I breezed through sunlight and shadow. The flip side was that, at fifty miles an hour, chilly air felt downright frigid, and sometimes, the only way to get out of a rainstorm was to keep riding. I got tough because there was no other option. But my hardiness was put to the test that first week of December. The hours I spent at the barn with Daisy, watching and waiting, left me brittle with cold. My bones felt like icicles. When I yawned, the frozen air caught in my throat and burned my lungs.

Baby hung out in the corral next to the barn. He was bonded to Daisy and liked being close to her. One afternoon, after checking on Daisy and chopping ice from the water tank with an ax, I crossed to Baby to brush him where he lay. At nine months old, Baby weighed close to a thousand pounds. Twin puffs of steam billowed from his nostrils with every exhale, like a dragon or an angry cartoon bull. The moisture in the air had frozen into tiny, glittering shards that sparkled around us. Baby was perfectly content to lounge in the snow, thanks to his substantial layer of fat and quarter-inch-thick skin. The black curls of his winter coat soaked up the December sunlight, catalyzing it into heat. Desperate for warmth, I crouched in the snow beside Baby and nestled against the slope of his shoulder. Baby was unbothered by my cuddling, and the heat of his body thawed my chest where I pressed against him. I wanted more contact with his soft, muscled warmth. I stood and rubbed his shoulder blades, then tentatively swung my leg across his wide body and sat on his back, straddling him. His body heat rushed through my jeans, through my skin, permeating the muscles of my legs, warming me to my very bones. I stopped shivering and took a slow breath. Baby glanced back at me but made no move to get up. His body beneath me felt like an ocean wave on pause—dense, yet yielding; powerful, yet soothing. He was broader than a horse, and more comfortable, cushiony and plush. I folded forward, circled my arms around his neck, and stretched my body out on top of his. My cheek rested in the curls of his coat, my legs draped across his flanks. His breath rocked me gently as I sank into the aura of his warmth. Baby remained relaxed beneath me, accepting me, as I sprawled across his back, as casually as I accept my cat into my lap.

From that day forth, I climbed on Baby's back whenever he happened to be lying in the corral when I checked on Daisy. One morning, I was stretched out on top of him when he began to rearrange himself beneath me, getting ready to stand. Instead of sliding off his back before he got up, I sat and gripped his barrel body with my knees. Several inches of fluffy snow blanketed the corral; if he threw me, the

landing would be soft. Baby lurched to his feet and ambled over to the water tank for a drink. I sat behind the mounds of his shoulders, giddy and grinning, my legs dangling at his sides. I scratched him between his shoulder blades while he drank. When he was done, I slid down the curve of his ribs, hugged his shaggy face goodbye, and, since Daisy still showed no signs of calving, walked back home.

A week or two into my vigil, Baby was standing and eating hay when I arrived at the corral. Snow drifted down as if on a whim, in no hurry to reach the ground. After peeking in on Daisy, I leaned against Baby's side for a quick hit of heat. Baby dropped to his front knees.

With his hindquarters in the air, his chin in the snow, he was kneeling. Laughter burst from my nose. I dusted the snowflakes from his coat and stepped across his back. He stood, lifting me with him, and I bent forward and hugged his neck. Baby went back to eating, and when he finished his hay, he wandered out into the pasture adjacent to the corral with me perched on his back. He walked slowly, shuffling through the snow, and I talked to him as we wandered. The UPS truck drove by on the county road, then stopped, backed up, and idled as we meandered past, the driver staring. When I posted a story online about Baby kneeling for me, a reader suggested that, since he knelt, he deserved the title Sir Baby. I sent her a lock of Sir Baby's hair, and the name stuck.

Bulls have been given the reputation of villain, monster, evil beast with a terrible temper, and it's unfair. It maintains a certain mythology. Bulls fight each other for breeding rights, as males of so many species do, but during the rest of the year, when they are not with the cows or once the cows have all been bred, bulls live together harmoniously. After seeing how much Daisy and Sir Baby enjoyed being brushed, I began brushing Mike's bulls, who lived across the road from the corral, to keep myself warm and entertained while staying close to Daisy. They showed their appreciation by allowing me to lie on their backs while they lounged around, though Sir Baby was the only one I rode. All the bulls I have known have been shy, sweet, gentle, and dear. When one or two of them decide to walk through a fence to find a haystack in which to bury their enormous heads (for fences are merely suggestions where bulls are concerned), all I need to do, upon discovering their transgression, is sidle between their noses and the hay, and this encourages them to lumber back to their pasture. They have the power to toss me out of their way like a rag doll in order to get more hay, but they choose not to. I doubt they even consider it.

It's not that bulls aren't dangerous, it's that there's a difference between dangerous and mean. All bovines can be dangerous because of their size and strength—far more people are killed each year by cattle than by sharks. Once, I was standing behind Daisy when she swatted a fly with her tail. Her tail struck me in the face, purely by accident, and the pain was so shocking, I thought I would end up with a black eye. I considered Daisy as potentially dangerous as the bulls, and the bulls as inherently lovable as Daisy.

Daisy's udder continued to grow, inflating in every direction, bulging out between her hind legs. It grew to boggling proportions, until there was no denying that the dairy cow is a mutation designed by man, achieved by generations of selective breeding to reap enormous supplies of milk from a single animal. Two weeks after I moved Daisy into the barn, the back wrinkle in her udder filled out. I was certain she would calve at any moment and I didn't want to leave her. Mike had an old camp trailer he had been given in trade for a favor, which he didn't particularly want or have use for, and which had been parked by the barn ever since. I was tired of running back and forth between my cabin and the corral countless times each day, so I moved into the trailer to be closer to Daisy, returning to the cabin only to feed myself and Charlie, Chloe, and Eli. I ran an extension cord between the corral and the camp trailer, which kept my laptop charged and powered a lamp by which to work, an electric kettle for tea, and a tiny space heater. The trailer had a platform bed, and I layered my sleeping bag inside Mike's heavy bedroll and piled blankets on top. Winter camping with a pregnant cow was less romantic than it might sound. I wore fingerless gloves to keep my hands warm enough to type, I slept in a wool sweater and hat, I swam through a haze of sleep deprivation brought on by waking several times each night to check on Daisy. The cold snap broke and the temperature rose to lows of 12°F, a glorious relief—a full thirty-two degrees warmer than the -20°F nights of the previous week, as profound a change as the difference between 48°F and 80°F. But it was still far too cold to be safe for a wet newborn calf.

In the darkness of 5:00 a.m., halfway through the third week of December, I crept out of the trailer to check on Daisy. I turned on the light in the barn, and in the soft, golden glow, I saw Daisy standing in the straw, staring at the wall, her eyes unfocused, her attention inward. Her tail was stiff and she held it straight out behind her, parallel to the ground. She mooed softly and swung her head around to lick her sides. She was in labor. Adrenaline surged through my body, tendrils of sleepiness vanished, and I no longer noticed the cold. I crouched in a corner of the barn, trying to make myself as small and inconspicuous as possible. Daisy circled and sniffed the straw, then lay down on her side.

When she pushed, her top hooves lifted slightly off the ground from the effort. She met my eyes, inhaled deeply, and pushed again. Two small, glossy hooves emerged from her body. She rested for a moment, then pushed again. The calf's delicate ankles revealed themselves, and then its nose, nestled between its forelegs. Daisy paused to rest, murmuring quiet moos as if she were speaking to her baby as it journeyed from her body into this world. She pushed again, and as the calf's head and shoulders appeared, I darted over to wipe a bit of membrane from the amniotic sack away from the calf's nostrils. With one last push, the calf's slick, black body slithered out into the straw.

Daisy lurched up and, standing over her new baby, began slowly, rhythmically licking off the slippery amniotic goo that coated his body. The calf lifted his head and snorted and shook his ears. His coat was black, edged with silver. At the center of his forehead spun a whorl of hair like a fingerprint. The joints of his knees were huge, foreshadowing, like a puppy's paws, his future stature. The calf fumbled his gangly legs beneath him, trying to stand. He heaved himself up and toppled over into the straw, shook his head, and tried again. He got his hind legs up and rested on his front knees for a moment, then stood, wobbling and unsteady. Daisy continued to lick his body and, with a great swipe of her tongue, knocked him off balance, and he fell down into the straw. My impulse, when watching a newborn calf trying to stand for the first time, is to rush to their side and hold them up, and I have to force myself to keep my distance, to allow them to figure it out on their own. But even with Daisy diligently licking him dry, the calf was shivering, and after a few attempts, he tired and lay sprawled in the straw.

I ran to the trailer, ripped the flannel pillowcase from my pillow, and joined Daisy in drying off the calf. As I rubbed his body with the flannel to fluff up his coat, Daisy lay down to rest beside him. The calf was still shivering, so I ran back to the trailer for a blanket and the little space heater. I plugged the heater in at the barn, sat beside the calf with the heater in my lap, and pulled the blanket over both of us like a tent. It did not take long for the calf to warm up in the dark pocket of heat, and soon he was trying to stand again beneath the blanket. I ran my hand through his coat. It was dry and fluffy. I wiggled my finger into his mouth. His mouth was warm and he started sucking on my finger, eager to nurse. I threw off the blanket and unplugged the heater, and returned to the corner of the barn as the calf stood, teetering on his little pointed hooves, and took a few tentative steps without falling. He bumped against Daisy's flank and she heaved herself up to standing. The calf wobbled to her side and began nosing around for a teat, ready for his first meal. I held my breath, on the edge of relief, for once a calf is up and warm and has had its first drink, the most precarious stage is past. As the calf wrapped his tongue around a teat and pulled it into his mouth, Daisy's hind leg shot to the side and she kicked him away. The calf stumbled backward and fell into the straw. I leaped up with a hot flash of anger and ran to the calf to make sure he was not hurt. I hadn't expected Daisy to have trouble nursing after she had adopted Baby and shown him such devotion. But as quickly as the anger rose in my chest, I felt ashamed by my reaction. At the dairy, Daisy's calves had been taken from her immediately after birth. This was a new experience for her, and though it was natural, it was unfamiliar. So often, our conditioning overrides our instincts. So often, the unfamiliar feels wrong, no matter how right it might be. I couldn't blame her for kicking her calf away. But I could help her, remind her of the familiar while introducing the new.

Before the calf had a chance to regain his footing, Daisy lay down and delivered her placenta. She stood immediately and began gobbling it up. It might seem strange that an herbivore would choose to eat this blood-rich organ, but it is an innate impulse in nearly all cows—perhaps to protect their newborn, for in the wild, the placenta would attract predators; or perhaps the cow receives vital nutrition from her placenta after the extraordinary feat of birth. Perhaps it is both. I ran to the trailer and grabbed the carafe of my electric kettle. The first flush of dawn was spreading across the eastern sky. While Daisy devoured her placenta, I sat beside her and began to milk into the carafe. She produced far more colostrum—the vital first milk that provides immune boosters to newborns—than her calf could consume, and I planned to freeze the extra for future orphans who weren't able to get colostrum from their mothers. Daisy ignored me, unfazed by my milking. Her calf stood behind me, nosing my back and sides in search of a teat. I heard a truck pull up outside the barn, and Mike appeared in the doorway.

"She did it!" he said with a smile. Daisy glanced at Mike but remained focused on eating her placenta.

"Yes! And your timing is perfect," I said, looking up from Daisy's side. "She kicked at the baby when he tried to nurse. Will you help him find a teat while I keep milking?"

Mike stood behind Daisy's calf and guided him to Daisy's udder beside me. The calf latched onto Daisy's back teat and began slurping and suckling as I milked her front teat. After Daisy finished her placenta, she remained rooted and still as I milked her with her calf by my side. After a few minutes, I stopped milking and slowly crept away as Daisy's calf continued to nurse. I leaned against Mike near the back of the barn, anxious to see Daisy's next move. Daisy swung her head around to look back at her calf, mooed softly as she had during labor, and stood calmly until he finished his first meal.

When her calf stopped nursing, Daisy turned to lick his back and neck. The calf, warm and fed, pranced in circles around Daisy, who turned in tight circles herself, trying to keep up to keep licking him. The morning was mild and the sun grew strong and bright, the day on its way to becoming the warmest of the month. I opened the gate between the barn and the corral, and the calf raced out into the snow, bucking with the pure exuberance of being. He trotted back to Daisy and she wrapped her neck around his body. He nuzzled his face into her cheek. When the calf plopped down in the sun to rest, Daisy lay down behind him, her body spooning his protectively. Though the newborn weighed about eighty pounds, he looked tiny next to Daisy, curled up at her heart.

Mike was on his way to feed his cows, and I rode along with him, leaving Daisy and her calf together to bond. I had more milking to do; I had straw and afterbirth in my hair; I had a shower to take, work and sleep to catch up on. But for now, with Mike at the wheel, I could let my mind and body melt. There would be time for the rest. I sank into the seat of Mike's truck with a smile and a sigh. There are moments, precious because they are fleeting, that feel so infused with harmony and hope and safety that there is room for nothing else. In those moments, I forget that they are fleeting. In those moments, it feels like they will last forever, that life can be no other way. As we drove to the haystack, one of my favorite songs that I hardly ever hear came on the radio, "San Francisco" by Scott McKenzie. By the second verse, I knew the calf's name was Frisco. The name conjured an image of a burly guy with an anchor tattoo and a heart of gold. I had only known Daisy's calf for a morning, but I believed the name would fit him perfectly.

CHAPTER 3

Communicating with Cattle

n Wyoming, the passage from winter to spring is as dramatic as the dawn; rebirth on a grand scale. Warm wind melts the snow so quickly I never have the chance to say a proper goodbye. The creeks rise with snowmelt, the earth begins to thaw. The air is sweet and mild. After winter's intensity, this softness in the lungs can easily be mistaken for a promise of calm, yet life is anything but calm. It's fitting that the name of the season is also a verb, for spring is *action*. Dormant vitality bursts forth into form. What was unseen emerges—blades of grass rise from the mud in vibrant ribbons of life force glowing green in the sun, the color startling after winter's palette. Leaves unfurl, babies slide from their mothers' wombs. Dawn's glow across the eastern sky arrives earlier with each passing day, and twilight lingers long after sunset, the land illuminated by rose-colored light. The rhythm of our work mirrored this acceleration—there was so much to do and more daylight in which to do it. Slow evenings beside the fire vanished as abruptly as the snow. Mike and I checked on heifers throughout the night, alternating shifts so we didn't get too sleep deprived, and days were spent calving, fixing fences and waterlines, prepping pastures, clearing weeds from the garden, and addressing the half-finished projects from autumn that had been buried by snow for months.

Since I worked from home, I kept an eye on Mike's cows while he was at work. When a cow was ready to calve, she wandered from the herd and found a secluded spot to have her calf—at the top of a hill, where she could see intruders approaching from a distance; sheltered against a high bank; or completely hidden at the bottom of a draw. Much of the work of calving entailed sitting in the sagebrush with binoculars, making sure the calf was not breech and that calving went smoothly, watching the mother cow lick and nuzzle her newborn, quietly waiting for the calf to stand and suckle. This didn't feel like work, like a task to be completed as quickly as possible. It felt like being included in a tender secret.

After weeks of gorgeous spring days and mild nights, the snow long melted and the mud dried, I woke, one April morning, to an incongruous quiet—the hush that comes with heavy snow. Out the window were dunes of snowdrifts, the aftermath of a predawn blizzard. That I slept through the roaring wind was a testament to the chronic exhaustion that came with the season. I threw on clothes, immediately concerned for a calf that had been born the previous afternoon. The day had been sunny and calm, the delivery smooth, the mother attentive, and the calf was up and nursing well before sundown. Young calves—even those just a few days old—are shockingly hardy. Their fuzzy coats insulate them from the cold, and their mothers' milk warms them from the inside. But a calf just hours old is much more vulnerable. Wind had whipped the snow into knee-deep drifts that could bury a newborn. I shouldered my door open against the banked snow to go searching for the calf.

Outside, I heard the crunching of tires on snow. Mike was driving slowly down the driveway, using the front of his truck like a plow. When he reached the cabin, I hopped in the passenger seat.

"What a lovely spring day," he said as I got in.

"This is wild," I said. "Did you know it was coming?"

He shook his head. "It blew in outta nowhere."

Mike was on the same mission as I—to find the newborn calf. The cows were huddled together at the corral and we waded through the snow, counting calves, looking everyone over, making sure they were all okay. Daisy's coat camouflaged her in the dimensionless expanse of snow. Her brown eyes, and Frisco by her side, were all that gave her away. The newborn calf was what we call a black baldie—she had a black body and a white face. We found her buried in a snowdrift with just her head poking out, nearly impossible to see but for the black ovals that marked her eyes like a panda's, beacons in the abyss of white.

Mike lifted the calf from the snow and carried her to the truck. I jumped in and he set her on my lap. The snow had soaked through her coat and chilled her to her core. I cupped her hooves in my hands and they were like blocks of ice. The inside of her mouth was cold and her tongue was limp. Mike drove us to my cabin and carried the calf inside, laying her on the floor beside my woodstove. "You OK with this?" he asked. I knew he had to get to work, and he knew I loved cabin calves. "Of course," I said. "Drive safe." I dried the calf with towels, rubbing her back and legs, her belly, her neck. I spread an old quilt on the floor and dragged the calf onto it, then folded the fabric around her, swaddling her entire body except for her eyes and nose. I knocked snow off one of my last pitchy logs and threw it in the woodstove, opening the air vents to get the fire roaring. I took off my coat and laid it on top of the swaddled calf. She was calm, but she didn't move at all, didn't moo, didn't even try to lift her head. I tucked her swaddled hooves under the woodstove. When Charlie was a pup, he liked to sprawl beneath the belly of the woodstove, so I knew it didn't get too hot under there. I sat beside the calf, stroking her forehead, talking to her as she warmed up. After a while, I un-wrapped her from the quilt, flipped her over, and reswaddled her so her other side was exposed to the heat of the fire. Then, since the cabin was a single room, I worked at my computer while watching over her.

About an hour later, I reached my hand into the quilt and found the calf's belly warm and dry. I tucked my finger into her mouth and it, too, was warm. She sucked lightly on my finger. I heated a jar of

Daisy's milk and poured it in a calf bottle. When I untucked the calf, she made no move to get up. I stood above her and, with my fingers laced beneath her rib cage, lifted her little body to encourage her to stand, but her legs were like noodles and crumpled beneath her as I gently lowered her down. She drank just a few sips from the bottle while lying with her head in my lap.

This wasn't the first calf I'd cared for in my cabin, but the others had been preemptive saves—calves born too close to nightfall in terrible weather, likely to freeze to death overnight if left outside; or twins who'd been rejected by their mothers and needed special attention and care. This was the first baby who had been truly compromised before arriving in my house. She lay on her side with her cheek resting on a pillow and remained still, unmoving, all morning. By noon, she hadn't raised her head, one of the most natural acts for a healthy calf. Her breathing was rapid and shallow—too rapid, a desperate panting. I was confronted with the very real possibility that she would expire before she was able to regain her strength—the very act of trying to survive was depleting what little energy she had left.

I had learned to pay attention to respiration rates while training to be an EMT. Our county's ambulance service was staffed entirely by volunteers, and the certification course—nearly two hundred hours of classes and practical training—was free for anyone interested. When the training director had approached me at the gas station and asked if I might want to become an EMT, I replied with a wholehearted "Yes!" then and there. With a volunteer crew, emergency calls were dispatched to pagers and EMTs responded when they could. Wyoming is the least populated state in the country—though it ranks tenth in terms of size—and in rural areas, people are scattered like wildflowers. The more volunteer EMTs that were similarly dispersed across the county, the greater the chance of someone arriving quickly enough to make a difference. We were an amalgamation of artists, schoolteachers, secretaries, stay-at-home moms, and retirees, but the exams we passed were equivalent to those of any professional ambulance service.

I knelt beside the calf. On the ambulance, we would give oxygen to a patient with these symptoms. I picked up the phone and called my ambulance director. "This is gonna sound a little crazy," I said, "but I have a calf that was born yesterday and got buried in a snowdrift, and she's not doing well, and I wondered if I could borrow some oxygen from the ambulance?" He boomed a laugh from the other end of the line. "Sure," he said. "Take one of the spare tanks." I thanked him profusely and grabbed my coat.

The clouds had lifted. The sky was a vibrant blue, the sun bright, and the snow had already melted off my truck. The ambulance bay was just five miles from home, and there I gathered an oxygen tank, a regulator valve, and a non-rebreather mask. On my way back through town, I saw Mike at the gas station. I stopped to tell him the status of the calf, and he followed me back to the cabin. The calf was still lying on the floor where I'd left her. I set up the regulator and connected the mask, then turned on the oxygen and held the mask over the calf's little nose. Her eyelashes began to flutter, and her eyes opened. She inhaled deeply. She seemed to drink in the oxygen. Her breathing slowed until it was just half the rate it had been. Mike and I sat transfixed as the calf came to life, blinking, wiggling her ears, moving her legs. I laughed in astonishment at how quickly she transformed before our eyes. I drew the mask a few inches away from her nose and she lifted her head to follow it. After fifteen minutes, she lurched to her feet, wobbly but strong, and took dainty steps around my cabin. The oxygen had done good work. It was time to take her to her mother for another boost of life force—mother's milk.

Mike walked the calf's mother to the corral and put her in the little barn where Daisy had birthed Frisco. We drove the calf down to the corral and Mike carried her into the barn. When he set her down, the cow ran to her calf and the calf immediately latched on to a teat, clumsily drinking as the cow stood patiently. When the calf finished her meal, she curled up in the fresh straw we had laid out, and when I peeked in on them an hour later, the cow was lying beside her calf. We left them together in the barn. It was better for them both to stay together, rather than taking the calf back to my cabin overnight. I ran down to the barn at midnight and at 3:00 a.m. to make sure the calf was warm and doing well, prepared to bring her home with me if she seemed chilled or weak. When I went down at 6:00 a.m., the calf was standing and nursing, the surest sign of recovery one can have.

Mike did not own a ranch. His forty acres were not nearly enough grazing land for his entire herd. During the winter, when snow blanketed the valley, his cows lived at home and we fed them hay every day. After calving season, we trailed his cattle to larger pastures that Mike leased from landowners. Mike's spring pasture lease was a thousand acres of native grass meadows. This land was untilled and unirrigated, absent of any sign of human intervention other than the barbed-wire fences that bordered the perimeter, separating neighbors. From the pasture, there were no telephone poles, no buildings, no rooftops in sight; there wasn't even cell service. A creek meandered through the pasture, supplying water for cows and wildlife. Cottonwood trees and willows lined the banks of the creek, providing shade. Perennial grasses swept across the gentle hills, peppered with sweet clover and bright shocks of Indian paintbrush. Look closer and one might find wild onion, beetles and lizards, mushrooms after a rain. Purple-gray belemnites—distant relatives of squid—lay cached in the dirt, shaped like bullets and fossilized into stone, relics from 150 million years ago when this area was a great inland sea. Dark mouths of dens gaped from the sloping draws that edged the grassy flats, home to rabbit and coyote. Nature's symphony filled the air— meadowlarks and blackbirds warbling from juniper perches; insects on missions to pollinate and procreate; cottonwood leaves whispering in the breeze.

For much of the year, cows and bulls live separately from one another. Cows are fertile year-round, and to ensure that calves were born in the spring when the weather is milder—April blizzards not withstanding—Mike's bulls lived in a bachelor group until they joined the cows and heifers. About a month after we trailed Mike's cows and calves to spring pasture, we hauled his bulls by horse trailer to join them. This time, Sir Baby was among their ranks.

I went to visit Baby often. If I didn't spot him right away, I wandered the pasture until I found him lounging in the shade chewing his cud or saw him across the creek nuzzling necks with a cow. On one such traverse, I found myself face to face with a baby fox. We stood frozen and silent, our eyes locked, staring at each other. Minutes went by—how many, I do not know. I have a poor concept of time as it is, which disintegrates completely when I'm in the presence of a baby fox. Eventually, I glanced away to see if the mother fox was nearby, and when I looked back a millisecond later, the little one had vanished. Sometimes, I brushed Sir Baby with a curry comb I brought with me; other times, I just watched him from afar. I wondered what it must feel like to be a bull—to be home in a body built of two thousand pounds of muscle and bone, to carry that colossal head, to strike the earth with four walloping hooves. For all their mass, bulls move so gracefully. Sir Baby glided across the pasture on his way to the creek, fluid and nimble, covering ground with surprising speed.

A few weeks in, I couldn't find Sir Baby. On such a large pasture with hills and trees, it's impossible to see all the cattle all the time, so I chalked it up to bad timing. I didn't see him the next day, either. The third day, I asked Mike to ride out with me to help me look. "I'm probably being paranoid," I said, "but I'm worried." We crisscrossed the pasture on Mike's four-wheeler, but never caught a glimpse of Sir Baby. We returned the fourth day to walk the creek. There were a few treacherous spots along the creek where the banks were high and steep, and the water deep, where a cow could get trapped and die. It was rare, but Mike had lost two cows to these traps in previous years. We set out afoot at opposite ends of the creek. There was no way for us to communicate since there was no cell service, and our plan was to walk until we met each other somewhere in the middle.

I picked my way along the water's edge as quickly as I could, channeling my anxiety into action. Light flickered across my path through the leaves above in a dance of light and shadow, then faded as a cloud coasted across the sun on a high breeze. I walked until I thought I would surely see Mike around the next bend. And then I saw Sir Baby. He was in the creek, standing knee deep in the cold water. Relief surged through my body when I saw he was not in a trap. A large depression in the soft, gently sloping bank showed where he had been resting. I splashed out to Baby, hugged his neck and kissed his face. When I walked up the bank toward the meadow, Baby watched me from the water, refusing to follow me. I returned to his side and stood with him under the willows, rubbing his shoulder blades, and waited for Mike to find us.

When Mike arrived, we gathered leafy branches from the bank

and shook them behind Baby to urge him out of the water. Baby took halting steps, limping toward the bank. He could barely walk. As he emerged from the creek, we saw a gash across the ankle of his left hind leg, just above his hoof. His lower leg was swollen, his flesh raw. An area the size of my palm had been sliced open, likely by barbwire. Perhaps Baby and a neighbor's cow had fallen in lust from opposite sides of the fence, and this was the consequence. Fences are a human construct, and we seem to be the only species on the planet to care, deeply, about what they represent. Every other manifestation of nature—seeds, water, cattle, insects, wind—answers to a higher power.

Mike drove home, returned with the horse trailer, and backed it into the meadow as close to the creek bank as he could. We flanked Sir Baby, slowly guiding him up the bank and across the meadow to the open door of the horse trailer. I was grateful Baby was tame and trusting, for he stepped into the trailer without balking. Once we got Sir Baby home and settled in the corral, I called my vet and learned he was out of the country.

I didn't feel comfortable giving Baby shots of antibiotics without direction. The dosing schedule can vary by degree of injury, and I didn't have the experience to determine how much to give him. But I needed to work on his open wound and the infection that had swollen his lower leg. Based on what I'd learned about charcoal from my EMT training—it can be given orally in certain poisoning cases because it adsorbs toxins—I thought perhaps a charcoal poultice could help draw out the infection in Sir Baby's leg. I built a small fire on the ground outside my cabin with the detritus from my woodpile. The bits of wood burned through quickly and transformed to sticks of charcoal. I put the chunks of charcoal into a dish and ground them into powder with a large rock, then mixed in water to make a paste. I ran down to the corral. Baby was lying down, and I sat beside him and carefully spread the charcoal paste over the gash on his ankle. When the poultice dried, it looked like a neat coat of tar, slick and black and shiny, and formed a protective layer that kept dirt out of the wound while allowing it to breathe and drain. Over the course of twenty-four hours, the charcoal puffed and plumped as it drew out the infection, and fell off in chunks coated with a thick layer of pus. Each day, I picked off the remaining bits of saturated charcoal from the day before, cleaned the wound, and applied a fresh coat of charcoal to Baby's ankle.

By late June, the days had become sweltering and rainfall scarce. The unirrigated native grass in the valley, once velvet green and lush, began to dry out. It was time to trail Mike's cattle to their summer pasture on the mountain, where they grazed until snow arrived in autumn. We gathered Mike's herd from their spring pasture and trailed them home for the night. The next morning, Mike and I saddled our horses at first light and set out with the herd while the air was still cool from night. We headed out the back gate on deer trails through Bureau of Land Management (BLM) public land, planning to gain a thousand feet of elevation before the sun got high and the valley got hot.

The cows knew the way to the mountain pasture and were eager to go. They knew fresh grass and cooler temperatures awaited them. The only reason Mike and I ever needed to tag along was to open gates along the way and make sure the calves stayed with their mothers during the day-long meander up the mountain. We had about a mile of BLM to cross to reach a dirt road that led to the mountain. The cows headed out on the trail in a neat line, their pace calm but determined through the dry cheatgrass dotted with sagebrush and yucca. But within a quarter mile, the cows noticed the yuccas were flowering. Yuccas are one of the last of the high desert plants to bloom. After the Indian paintbrush and lupines flower and fade, yucca plants send up a single scaly stalk, two to three feet tall, from the center of their spiky fronds. The scales fold away from the stem and fatten into buds, then burst open into pale yellow bells—forty to sixty flowers, each the size of a golf ball, spiraling down the stalk. Tiny white moths pollinate the flowers, and soon after, sweet orbs ripen where each flower had been. Yucca flowers and fruit are like candy to deer and cattle. Instead of trickling toward the mountain in an orderly line, the cows scattered across the BLM in search of yucca flowers. They trotted from plant to plant, udders swinging, nipping off flowers and devouring them by the mouthful.

Mike and I raced back and forth and in circles chasing the wayward cows, trying to round up the fractured herd to get them back on the trail. The cows paid us no attention. They head-butted each other away from yuccas they intended to claim and were impossible to move once they'd latched onto a stalk laden with flowers. By the time we managed to bunch part of the herd together from behind, the lead cows had dispersed again, on the hunt for more flowers. Heat rippled across the cheatgrass as the sun floated higher in the sky, the temperature rising with it. The mile of BLM, which should have taken about twenty minutes to cross, took us three hours. Once we hit the dirt road and the yuccas were behind us, the cows strung out in a line and ambled up the mountain.

The next day, it was 100°F by 10:00 a.m. I looked out the window, cursing the heat, and saw a lone black cow standing at the back fence. We must have missed her when we rounded up the cows after their yucca rampage and accidentally left her behind. I hiked out and opened the back gate, knowing she would be thirsty. I circled behind the cow to walk her to the corral since she belonged on the mountain and would need to be driven up. The cow snorted at me and ran back into the BLM. I chased her briefly, but it was too hot to try to herd an unwilling cow on foot by myself, so I left the gate open for her and went back to the cabin. I knew she'd come in eventually to get water. I had her held in the corral by the time Mike got home.

The cow's udder was full and tight, a telltale sign that her calf had trailed up with the herd and was on the mountain without her. We loaded her in the horse trailer and Mike drove her to the very top of the mountain pasture. From there, she would have to walk through all the other cows and calves to get to the mountain spring for water, our hope being that in doing so, she'd find her missing calf.

The next morning, she was standing at the back gate again.

I couldn't believe it. She'd walked all the way down the mountain, by herself, in the dark, somehow crossing through three locked gates, and come back home. I went out to herd her in, and once again, she ran off when I circled behind her. I left the gate open and left her alone. When Mike got home, the cow was still out in the BLM just beyond the fence. He rode out on his four-wheeler to bring her in so we could drive her back up the mountain to try once again to unite her with her calf. As he rode a wide, slow circle behind the cow, Mike spotted a tiny black calf nestled beneath a sagebrush. The cow hadn't been separated from her calf during the yucca frenzy, she'd gone off alone to *birth* her calf. Her udder was so tight because she had just calved. By taking her up the mountain, we'd inadvertently separated her from her newborn, and she hoofed it, quite literally, down a mountainside in the dark to rectify our mistake. Mike scooped up the calf and held her on his lap as he maneuvered the four-wheeler in through the gate. The cow followed her calf and came in behind him. They would be spending the summer at home.

Along with Daisy and Frisco, and Sir Baby since he was recuperating, a small group of cattle lived year-round on Mike's property. Mike's "Special Project" cows were a motley menagerie who, for various reasons, could not take the long walks to seasonal pastures or needed extra attention and care—the old, the injured, the orphaned. Unlike other ranchers, Mike kept his cows until they died of old age. Most ranchers sell their mother cows when they reach eight to ten years old, and these culled older cows are replaced in the rancher's herd by young heifers. From a business standpoint, this makes perfect sense. While an eight-year-old cow is likely to be strong and healthy, as cows age into the teens, their fertility can drop off. And, like elder humans, elder cows tend to require more care: they need special consideration, nutritional supplements, more of the rancher's time. When the cow isn't having calves, there's no income, via calves to sell, to balance the expenses that come with care and feeding.

But little about Mike's approach to ranching was conventional. Mike started his cow herd after his daughter died. He bought ten ten-year-old cows that were on their way to slaughter and having them to tend and feed made him get out of bed every day. Over the years, his herd grew from ten cows to close to a hundred, all of them descendants of the original ten. Mike's cows had helped him through overwhelming grief, and he believed his elderly cows deserved to retire in peace and comfort, to live out their days at home in the family herd of their daughters and granddaughters. This isn't a cost-efficient way to run a ranch, because it's part animal sanctuary, but Mike offset the expenses through his job outside of ranching. He always had a number of very old Grandmother cows at any given time.

The Grandmothers, Daisy and Frisco, the cow who had calved in the BLM, her calf who I named Star, Sir Baby, and the other Special Project cows summered at home. Sir Baby's wound healed remarkably quickly with daily applications of charcoal. The infection drained and the swelling went down so rapidly I never gave him antibiotics. He was soon walking without any hint of a limp. The Grandmothers were much slower. Their bones creaked audibly. They took tiny, careful steps, not unlike the way my grandmother walked from supermarket to car when she was nearly ninety years old. They traversed Mike's property with Daisy and the other Special Project cows, it just took them much longer to get from place to place, and they didn't roam as spontaneously as the younger cows. The Grandmothers often babysat a group of calves while the mothers wandered off to graze. They licked Star and Frisco, who soaked up the extra attention with their eyes closed and chins in the air. The Grandmothers had no biological reason to dotingly tend to another cow's calf. It was apparent their behavior stemmed from an instinctual understanding that the care of each member of the herd determined the well-being of the herd as a whole.

Mike's mean old goose, Ricardo, found this small group of cows less terrifying than the full herd, and better company than the chickens. He waddled after the Grandmothers and napped beside Frisco in the sun. Whenever I approached, perhaps with a book, to lay against Frisco's shoulder, Ricardo hissed and honked and tried to flap me away. His allegiance to the cows was so complete, we renamed the Special Project cows the Goose Group.

Frisco was a gangly, gregarious darling. About a month after he was born, I had felt two smooth, symmetrical domes, not much larger than the diameter of a pencil eraser, hidden beneath the hair on his forehead. They sprouted into stubby knobs, which made his head appear rectangular and a bit Frankensteinian; then, over the course of months, they lengthened into adolescent horns. Frisco was a cross between a horse and a dog. He galloped like a horse, was quickly approaching the size of a horse, and bubbled over with the exuberant joy and camaraderie of a dog. In a way, and to degrees, this comparison is true of cattle in general—they learn their names, if they are lucky enough to be named, and come when called. They are affectionate and intelligent, loyal and intuitive, each with their own distinct personality.

One bright afternoon at the tail end of August, I realized I had left my sun hat at the corral when I'd milked Daisy that morning. Daisy saw me walking to the corral and set off in her determined stride behind me, leaving Frisco in the pasture behind the cabin with the other cows. When I noticed her following me, I waited for her to catch up and we walked to the corral together. I got my hat, then grabbed a curry comb from the barn. Daisy made my life so easy by following me, and so, to give her something in return, I began brushing her in the corral. As I brushed her, the light grew dim. The sky that had been blue and sparkling when we headed to the corral darkened ominously, lost all dimension. Wyoming is where Zeus has his tantrums. Weather comes in hard and fast—it can be brutal, but it's often short-lived. Sometimes all one can do is hunker down and wait for weather to blow on by. A storm was flying toward us, and I decided to stay at the corral until it passed rather than trying to beat it home.

The wind barreled in, filling the air with a cloud of red dirt from the road. Daisy turned her butt to the wind to shield her face, and the gusts blew her heavy tail horizontal against her flank, parallel to the ground. The wind accelerated, became so ferocious, it felt like being trapped under a wave. A shed beside the barn blew into pieces, wood flying. Full sheets of corrugated tin flew off the roof of the barn. Daisy galloped past me, heading to the open door of the barn, her head low to the ground to protect her eyes. I sprinted after her. A three-by-eight-foot sheet of roofing tin flew past my face, close enough I could see the screws rattling in their holes where they had been ripped from the wood frame of the barn. As we reached the barn, the gray cloud-bank that loomed above us unloaded its contents. It was like a hatch opened and hailstones the size of apricot pits plummeted to earth. The sound was like a freight train, the deluge so dense it was a curtain I could barely see through. Daisy and I stood in the barn, our eyes wide, our bodies rigid, bracing ourselves against the wailing chaos outside. I can so easily slide into feeling like I am in control—of myself, of my life, of my animals' well-being—and that feeling is much easier to live with than its opposite. But since I've been in Wyoming, Mother Nature reminds me, on a regular basis, that this feeling is an illusion.

Eventually, I took a deep breath, which made me realize I'd been holding my breath. When I took a deep breath, I recognized that my body knew, before my mind had caught on, that the worst of the storm had passed. The hail and wind descended upon us less violently. I saw a hint of light glowing behind the gray wall of clouds. I inhaled the storm-sweetened air, and as abruptly as it began, the hail ceased. It littered the earth but was already melting. The wind downshifted to a breeze. I stepped out of the barn and a rainbow blazed across the horizon, a gateway, an ephemeral gift. The sky brightened back to blue and the sun burst forth, the light golden, shadows sharp. The rainbow disappeared.

Tin and timber from the corral and barn were scattered across the hillside, over the hill, almost all the way to my cabin. I jogged after Daisy as she trotted past the cabin to find Frisco, bellowing urgent moos. When we reached the other cows, we found them casually grazing. The cattle were calm, and the cabin untouched, as if the chaos we'd witnessed had missed this side of the hill entirely. Just a few leafy branches were torn from the cottonwood trees. I collected the branches from the ground and offered them to Daisy. She nipped off the heart-shaped leaves one by one and ate them like chips. I left her with Frisco and the rest, and burned off some of the adrenaline in my bloodstream by gathering up the splintered wood and scattered tin. When I talked to Mike, he said the radio had called the storm a microburst, cousin to the tornado. Daisy seemed relaxed by dusk, but her body revealed how stressed she had been—for a week, her milk tasted so metallic, it was undrinkable.

In mid-October, clouds obscured our view of the mountain; when they dispersed, the mountain was white. The mountain had seen occasional snow flurries over the previous weeks, but this time, the clouds left more than a dusting in their wake. With the arrival of snow, the cows would be ready to trade their alpine meadows for fall pasture in the valley. Mike and I headed up on horseback to gather the herd and trail them down. The mountain was brisk and bright as we rode across the high meadows and past ancient granite outcroppings, descending out of the snow and pines, down to the junipers and red sandstone buttes of the valley where the conditions were dry and mild. The herd ambled peacefully, the experienced cows leading the way, the calves following their elders. With the breeze in my hair, I basked in the sun-warmed scent of my horse, the rhythm of his shoulders, the creaking of my saddle, the sound of hooves on dirt.

Once the cattle were off the mountain, we weaned the calves. The calves were old enough to no longer need milk—in addition to nursing, they had started grazing of their own accord—and weaning allowed the mother cows to dry off before winter. I sorted Mike's herd myself, separating cows and calves into different sections of the corral, because I could do it much faster than Mike. I had become intimately familiar with cattle through my daily interactions with Daisy, Sir Baby, and Frisco; and I had learned from the yucca fiasco that it's much easier to work with bovines on their terms rather than trying to force them to conform to human logic. But when it came to communicating with cattle, the one who helped me most was Charlie. Charlie was tame to Mike and to me, but he was moodier than any domestic dog I've known, and sometimes, he expressed his fear or annoyance by snapping. Purely out of self-protection, I learned to read his subtle nonverbal cues—the position of his whiskers, the dilation of his pupils, his muscle tone and microexpressions—to determine his mercurial moods and predict his actions. It's a skill I learned with practice and attention, one that I believe anyone can learn. My ability to notice and interpret the minute fluctuations of Charlie's body translated to my interactions with cattle, without me even realizing it at first. I could read their silent declarations, understand the messages they broadcasted through slight changes in their posture, the position of their ears, the direction they moved their eyes or shifted their weight. I had a knack for sorting cattle because I could read their body language to predict their movements, and I used my body—sending signals through my shoulders, hips, and fingertips—to communicate with the cows and calves in their own language.

Once the cows and calves were sorted, we practiced low-stress weaning. We kept the calves in the corral and the cows in the pasture adjacent to the corral, separated by just a rail fence. Since the cows and calves could see, smell, and communicate with each other, they remained calm and quiet during this period of weaning. They did not bawl desperately for one another, as is the case when cows are abruptly separated from their calves by distance, and the calves did not get sick from stress. Over the course of a week, both cows and calves became accustomed to their independence. The mother cows drifted farther from the corral for longer stretches of time and the calves formed their own little gang. Then we moved the calves to a separate pasture while the cows dried off.

The worst part of having cows is when it's time to sell calves. Ranchers run the breeding stock—the cows and bulls—and every year, the calves are loaded into trailers and sold at auction. This annual calf sale is the primary, if not only, source of a rancher's income. After they're sold, the vast majority of calves are shipped to concentrated animal feeding operations, or CAFOs. CAFOs can each hold tens of thousands of cattle to upward of a hundred thousand cattle at any given time. In CAFOs, cattle are confined to pens where they are fed rations of corn, soy, and agricultural by-products. They are often given growth hormones, and preemptive antibiotic treatments are a standard practice to keep them from getting too sick too quickly from this unnatural diet and environment.

In my rural ranching community, thousands of calves were shipped off to CAFOs every autumn. As Mike prepared to sell his calves, I thought about what awaited them. After being born into a family herd and spending their days frolicking vast pastures, nurtured by their mothers and grandmothers; after grazing the native grass their bodies were designed for and on which they thrived; after being serenaded through the changing seasons by sandhill cranes and mourning doves in spring, meadowlarks and bluebirds in summer, owls and wild turkey in autumn; after all the care we gave them—they would be loaded into a semitrailer, the raging noise of the highway and bitter scent of panic from the animals crammed in beside them an unescapable assault on their senses; then unloaded into the desolate pens of a CAFO, barren and brown and barred; packed together with strangers in a strange land, with no room to roam, no trees, no birds, just bodies and dirt and shit; the final, intolerable months of their lives steeped in boredom and despair. Mike felt it was too great a betrayal to sell his older cows into this system. I didn't think any animals deserved such a fate. With the money I'd earned from my book—the nest egg I'd stashed in the bank—I started buying Mike's calves.

CHAPTER 4

Belonging to a Place

Daisy was pregnant, bred by Sir Baby. Daisy had nursed Baby, had raised him from a calf, but they were not related by blood. They were not precious about this development, and neither was I. My extended vigil preceding Frisco's birth had taught me what to look for—and what to wait for—as Daisy got closer to calving again. Her expanding udder was still startling as it grew to seemingly impossible proportions, but I waited patiently for the back wrinkle to fill out, and then I waited more, fluctuating between giddy anticipation and calm observation. The alchemy of spring rains and warm sun had given rise to lush grass, and the cows drifted about the pasture with their heads down, grazing the days away. One mild afternoon, a few days after the wrinkle in her udder had filled out, Daisy distanced herself from the herd, then set off alone toward the corral. No one followed her, not even Frisco. I watched her swaying body disappear around a hill, then trailed behind her, keeping my distance. I found her standing in the barn where she had birthed Frisco, uninterested in company, uninterested in food. Her attention was inward.

I shut the gates to the corral to keep the other cows from bothering her and filled the barn with fresh straw for Daisy to nest in. I piled hay in a corner, in case she got hungry, and left her by herself. I checked on Daisy periodically throughout the evening, but something in me knew her calf would come with the sunrise, and I didn't go down to the barn during the night. When I woke at dawn, I stuck to my usual morning rituals with Charlie and Chloe and coffee and the computer, in no rush to get to the corral, even though I was filled with an unshakable knowing that Daisy was, right then, in the process of birthing her calf. I had sensed, during Frisco's birth, that Daisy would have preferred to have been alone. As hard as it was to stay away, I wanted Daisy to calve in the manner she wished, in peace, without me. I was checking email when suddenly, something I can only describe as a telepathic command demanded I go to the barn. I pulled on my boots and trotted down the dirt path. As I neared the corral, I could see Daisy standing and blurry movement around her legs. With a few more strides, I glimpsed Daisy's newborn, swaying on splayed legs.

Daisy's calf was still completely wet and working on her balance when I got to the corral, which meant the birth had occurred just minutes before. Daisy was licking the calf's neck and worked methodically down the length of the newborn's body, drying and fluffing up her coat. Sir Baby's coat was solid black, and Daisy was solid white. Together, they had created a redhead. Daisy's calf was a bright, shimmering copper. Daisy gave her full attention to her calf, but she welcomed me, when I perched on my heels nearby, with a lick to my arm and shoulder every so often. When her calf had found her footing and was ready to nurse, Daisy stood patiently as the calf fumbled with her teats and had her first meal. I milked Daisy throughout the day, sitting beside her in the corral as she licked her calf and munched on hay, easing us both—and my hands in particular—into our new daily routine.

I didn't know if Daisy's daughter, whose genes were half Angus, would produce enough milk to be a dairy cow, but I felt I should name her as if she would be, just in case. A dairy cow's name requires special consideration: in addition to sounding nice when spoken, it must be easily hollered to call the cow in from across a field, and it must sound the same at a holler as it does at a murmur. "Daisy" sounds exactly the same in a speaking voice as when hollered. "Fawn," for example, would be a darling name, but it's difficult to holler and the whole sound of the word changes. The moment I saw Daisy's calf, I could feel her name in my mouth, but I couldn't figure out what it was. I posted a photo of her to my blog and asked readers for their suggestions. After a few days of reading through hundreds of names, I reached "Fiona," and it fit like a puzzle with what I felt in my mouth. Daisy, Frisco, and the strawberry calf were lounging in the pasture behind the cabin. I joined them, sat against a tree, and softly called out "Fiona . . ." Daisy's gorgeous calf rocked up from where she was resting in the sun and pranced over to me, nuzzled my neck, then took off, galloping to the edge of the pasture and back again, just because she could. She skidded to a stop at Frisco's shoulder and rubbed her forehead against his cheek.

Frisco had grown into a giant with a gentle spirit, reaching nearly seven feet from hooves to horns. His legs were so long, and his body so narrow, I joked that he was half Daisy, half moose. Sir Baby looked like a squat rhino beside him. Frisco was perhaps the only animal on the place without a job—even Eli took care of mice—but he was a lightworker, exceptionally friendly and social. He doted on Fiona. When the neighbors trailed their cows past Mike's land, Frisco ran to the fence to greet them, mooing high-pitched hellos as they trotted by. He brought me to belly laughs on a daily basis, chasing my truck up the driveway at a trot when I got home from a long day away, or lapping water straight from the hose with great swipes of his tongue. I adored him so much, I never begrudged him when he ate the lilac blossoms off every bush the day after they bloomed. He loved getting brushed as much as Daisy did, and all I had to do was wave the red curry comb above my head and Frisco would run to me. His hair was finer than any of the other bovines, jet black and soft as velvet. His glossy, obsidian horns were as thick as my forearms at the base. They curved in a crescent shape pointing heavenward and were just slightly asymmetrical—a jaunty crown.

had a complicated relationship with the internet. Through the internet, I was granted an avenue to share my work with the wider world without middlemen or gatekeepers. The freedom this allowed, and the connections this made possible, felt miraculous. A side effect I never expected was that a shocking number of people showed up at my house during their summer vacations. Each summer since my blog had gone viral brought an influx of unannounced visitors to my door, without warning, without invitation, without emails ahead of time asking if I had the time or the bandwidth for a visit. People in my community gave directions to my house when asked, because in a small town, that is considered being helpful. I had already been stalked for months by a reader, which had culminated in his arrest and conviction after he showed up with a loaded gun. I had no way to tell if the strangers parked at the bottom of my driveway, binoculars thrust in my direction, were friend or foe; if they just wanted a live glimpse of what they'd read about in my book or on my blog, or if they were plotting harm. I'm sure most, if not all, of the people who knocked on my door or took pictures from my driveway were perfectly wonderful individuals who just hadn't stopped to consider the effects of their actions and desires; who didn't think their unannounced appearances would be an imposition; who didn't realize each one felt like an invasion, disruptive at best and threatening at worst, and that I was always uncomfortable. It reached a point where I didn't want to leave my house, because I never knew if I would come home from errands to find groups of people peering in my windows; and I didn't want to be home, because home was no longer a place I could trust for peace and privacy. I was dreading the coming summer. I decided to escape. When Mike and I visited the owner of the mountain pasture Mike leased for the summer months, I asked if I could camp on his land for the season with the cows. He raised his eyebrows and nodded three times, a cowboy's blessing.

The mountain pasture was eight hundred acres of alpine meadows bordered by conifer woods. At an elevation of seventy-five hundred feet, it was nestled within ten thousand acres of private grazing land owned by a handful of other ranchers, a hidden sanctuary completely off the grid. I planned to live there until the snow kicked me off. I fixed up the old camp trailer, painted the cabinets, covered the seat cushions and water-stained ceiling with vintage fabric from the thrift store. The interior was a cozy seventy square feet, with a countertop and tiny gas stove along one side, a fold-up table and bench seats along the other, and a mattress filling the far end. I pulled the trailer next to the cabin while I worked on it so Charlie, Chloe, and Eli could get familiar with it and associate it with home, for they were coming with me, too.

The climate of any given year determined when the grass on the mountain was ready to welcome cows, and the grass dictated when we trailed cows up. The weather also determined when the roads were passable by vehicle, and whether or not they required four-wheel drive. When we trailed Mike's cows to the mountain, the dirt roads leading into the pasture were still too muddy to cross with the camp trailer, so

Daisy, Frisco, Fiona, and Sir Baby stayed in the valley until we could all move up together. After another week of sun dried the mud, I packed the camp trailer with everything I thought I would need for the next few months and Mike helped me move the whole crew to the mountain—Daisy, Frisco, Fiona, Sir Baby, Charlie, Chloe, and Eli.

Rather than trying to trail my four bovines such a long distance on a path they'd never been, with no experienced cows to help lead the way, we drove them up in the horse trailer. Daisy and Fiona loaded easily in the front compartment, but getting the two boys in the back was straight out of a silent movie. Sir Baby stepped into the trailer and, as Frisco was getting in behind him, Baby turned around and stepped out. Frisco climbed in, then, just as we were getting Baby back in behind him, Frisco turned around and hopped out. We went several rounds before giving up and closing the door behind Frisco. Mike said he would chauffeur Sir Baby to the mountain when he came to visit.

A rough dirt road led to the mountain pasture from the south, a stretch of which we used while trailing cows before veering off onto more direct footpaths. But in a vehicle, it was faster to take the highway up the mountain, through the national forest, and from there, weave down to the pasture on dirt roads from the north. Though, these were hardly roads—they were more like truck trails. The seven miles from pavement to pasture took an hour to drive. The route was narrow and bumpy, gouged by ruts, just two parallel tracks created by tires following the same path year by year across the land. Mud and snow made them impassable. Heavy rains could wash out sections, leaving deep trenches. In the unlikely event you met an oncoming vehicle, there was usually no room to pass, and one party would have to back up until the terrain flattened enough to allow one truck to pull off so the other could go by. Alpine meadows unfurled in every direction, exuberantly green, wildflowers scattered like jewels in the grass. Clusters of aspen blinked and swayed, their round leaves glittering like sequins. The reservoirs—spring-fed ponds that provided water to cattle and wildlife—shone in the sun, wild ducks paddling around the edges as eagles soared above.

Before ranchers had trucks with four-wheel drive and could check on their cows in an afternoon, it was a day's ride a'horseback just to reach the mountain pastures from the valley. Back then, cowboys stayed on the mountain with the cows. At the edge of a meadow, not far from the woods, sat an old cow camp—a weathered, one-room cabin built in the early 1900s; an outhouse that had toppled over decades prior; and a set of corrals built before the advent of the chainsaw, a fact Mike and I discovered when we dug up a rotting corral post to replace it. The post was the straight trunk of a pine tree, about eight inches in diameter, and the grooves in the buried end showed the tree had been chopped down by hand with an ax. That post, along with every other post and rail that made up the corrals, and the logs from which the cabin was built, had surely come from the woods to the west. The cabin was long neglected, contemplating collapse without yet committing. The windows had no glass, the logs of the walls were weathered and warped with great gaps between them, and one could watch the clouds through the holes in the sagging roof. Of the hundreds of acres I had access to, I chose to make my home base alongside the old cow camp. The corral was a perfect place to milk Daisy without interference from the other cows. The pre-dug outhouse was convenient, even though it had no walls and no seat, just two planks spanning the hole on which to balance. And it was close to water.

Not far from the camp was a spring box. Beneath a heavy wooden lid, a large hole, dug perhaps a century before, intersected an underground spring. The hole was about three feet square and lined, neatly and carefully, with stones. The water level was a few feet below the surface of the ground, and the natural spring flowed through this man-made cavern, so the water was always fresh and cold. The spring was my water source and the spring box my refrigerator, for it stayed about fifty degrees in the hole, even during the day. There was an interior ledge made of rocks just above the water line where I could store a few gallon jars of milk.

The cabin and the corrals were the only man-made structures I could see in any direction. The undulating curves of the native grass meadows rippled into the distance, studded with cows who spent their days grazing and lazing, edged with stands of aspen and pine. There were no utilities—no telephone poles, no electricity, no water or sewage lines, no artificial lights. I had a five-gallon propane tank hooked to the stove in the camper, since building fires would have been pointlessly reckless in a dry summer. When I needed to charge my phone, camera, and laptop batteries, I plugged them into a power inverter—a small box with two clamps wired into one end and an outlet on the other—which I connected to the terminals of my truck battery. There was no cell service at my camp or nearly anywhere on that part of the mountain. To use my phone, I walked about a mile to the west, through the woods, to a clearing at the edge of a sandstone cliff overlooking the valley. There, I had cell service. Anywhere else, the tethers of modern tech couldn't reach.

There's a common misconception about the land used for cattle grazing. Many people, unfamiliar with agriculture in general and the terrain specifically, believe grazing is a wasteful use of this land, that the land would serve humans better if it were transitioned to crops. But the mountain meadows used for grazing cattle and sheep can't be conventionally farmed. There's no irrigation and very little rain through the summer. Much of the land is difficult or impossible to access by vehicle. The growing season is incredibly short due to the elevation—snowstorms shroud the mountain in May, lingering snow-drifts stretch across north-facing hillsides in June, locals remember the years it has snowed on the mountain on the Fourth of July, and snow arrives again in September. I optimistically hauled three large, potted tomato plants up the mountain with me, and though they sat in direct sun and grew dozens of green orbs, the tomatoes never rip-ened. By the time they were killed by the first frost, the fruits held just the slightest tint of orange.

What does flourish on the mountain are robust native grasses. The vast majority of grazing land around the world is used for grazing precisely *because* that land will not support other crops. According to the 2017 USDA Census of Agriculture, less than 4 percent of all pastured land in the United States could be used for crops instead. Reasons for this include rocky or uneven terrain, thin topsoil, and lack of irrigation or rainfall. Instead of crops, these lands support hardy perennial grasses and the animals that graze them. Grazing land also includes land that was previously used for crops, but which has become so degraded through unsustainable farming practices that it can no longer be cultivated. When European settlers arrived on this continent, they cleared forests and drained wetlands not for cows, but for plows. While cattle and other grazers were often present on farms, crops were the priority. The native prairie, which once spanned the center of the continent and supported tens of millions of buffalo and other wild grazers with grasses that grew chest high, was tilled to plant crops. Combined with drought, the rampant tilling of native grasslands led to the Dust Bowl. Prioritizing land for crops continues to the present day, and grazing land consists of the leftovers and the in-betweens—land that never could, or can no longer, be used for crops; and land that hasn't yet been paved for subdivisions.

I was thrilled when Sir Baby made it up to our mountain home, but immediately noticed his left eye was bothering him. He kept it tightly closed and it was tearing up. Mike thought it was nothing more than dust from the trip, but the next morning, Baby still had his eye clamped shut. While Baby was lying down for his midmorning cud chew, I climbed on his back, scratched his shoulder blades, rubbed his neck, and moved my hands toward his cheek. Baby squeezed his eye closed even tighter as my fingertips approached it. I scooched my body up his neck until I was lying across his head and, with both hands, gently pried his eye open. I saw a flash of something pale and straight before he slammed it shut again. I sat back on Baby's shoulders and rubbed his ears, giving him some time, then sprawled across his head and opened his eye again. A huge grass seed was stuck in his eyeball, sharp side in. I moved down Baby's back so he could relax while I readied myself for the extraction. After another rubdown, I planted myself on Baby's head and pried his eye open once more, darted toward his eyeball with my fingertips, and quickly extracted the seed. Baby had his eye open by that afternoon and seemed completely unbothered by it the following day.

The animals were as delighted by our move as I was. Daisy, Fiona, Frisco, and Sir Baby had free range of the entire mountain pasture lease. Sir Baby roamed, breeding Mike's cows. Daisy, Fiona, and Frisco treated my camp as their home base, too. They grazed and explored, disappearing for hours, but returned often. Fiona frolicked with the other calves, straying far enough for long enough that Daisy bellowed booming moos demanding her return, with Frisco chiming in for good measure. High above the heat of the valley, the mountain was blissfully mild, with most days reaching the mid-seventies. The cattle were unmolested by flies, for there were few on the mountain, and I didn't see one mosquito the entire summer. A relief, on all counts, compared to the valley. I milked Daisy every morning, and her milk, fortified by the mountain meadow grass, was decadently creamy and as sweet as if it had been swirled with honey. I had plenty to share with Charlie, Chloe, and Eli, who gladly lapped it up.

I brought very little to the mountain with me. I had a dozen books and a few changes of clothes—enough to cover all possible weather and temperatures with little variety. I had my EMT kit, my milk pails, plenty of pencils and matches, and a potted thyme along with the ill-fated tomatoes. The bulk of my packing was nonperishable food: dried peas, lentils, and alfalfa seeds to sprout in mason jars for fresh greens; oats, rice, beans, and dried fruit; olive oil, balsamic vinegar, and honey; coffee and tea; a few seasonings; dog and cat food. Daisy's milk was a staple. When Mike came to visit, he brought eggs from his chickens and a package of frozen hamburger or the occasional steak. Once, he brought me a chocolate bar. I took an eager nibble, the base of my tongue curled, and I handed the bar back to Mike. The hyper-sweet blast of refined sugar felt like an assault on my taste buds after just a couple of weeks without it. How quickly we adapt.

I found the solitude absolutely divine. The only engine noise I heard came from commercial planes passing high overhead every few days, a neighboring rancher's vehicle when he checked on his cows every couple of weeks, and Mike's truck when he came to visit. The time I spent feeling lonely amounted to hours, not days. And I never felt alone—solo, but not alone. It was impossible to feel lonely while lying on the earth or beneath the explosion of stars. The stars were more spectacular than I'd ever seen them—brighter, crisper, infinite. I was surrounded by too much life to feel alone—trees and wildflowers, multicolored dragonflies, the miraculous and microscopic beings in the soil underfoot, all my animals, all the wild animals, all the birds.

After taking photos one morning, I popped into the trailer to drop off my camera and saw a commotion at the back window, a frantic fluttering. I thought it was an enormous bug and turned to reach for a glass to trap it so I could move it outside. My breath caught when the flutterer shifted position and revealed in silhouette the long, delicate beak of a hummingbird. I traded the glass for a wide-mouth half-gallon jar, thinking the least stressful maneuver for the hummingbird would be to move it like I move wasps and spiders. But each time I reached out with the jar, the hummingbird crouched at the bottom of the windowsill. It soon became obvious that method would not work without crushing the bird. I set down the jar and reached for the tiny beauty, cupping my hands around her. She was quiet and did not flutter as I carried her outside. She was as light as a palm full of sunshine. Standing in the grass, I opened my hands and the hummingbird sat for a moment in my palm, her throat glittering in the sunlight, and then, like a wink, she was gone.

But because I was the only human around for many miles, being present and focused, engaged in every moment, was a necessary safety measure. An accident alone on the mountain out of cell range would have been challenging at best. I remembered an ambulance call: a lone hunter had shot a bull elk, and as he prepared to field dress the elk, he slipped and fell toward it. A tine of the elk's antler punctured the hunter's femoral artery. He bled out, died on top of the elk. My every action, every step, had to be deliberate, made with care. Multi-tasking was just stupid. The mindfulness I eventually achieved—mindfulness that imbued every action, every moment—was a side effect of what the mountain required of me.

That deliberateness extended beyond myself, personally, to how I interacted with the land, the water, the sunlight, the night. I woke with the birds at dawn, and stopped working or reading when light faded from the sky. Time, the way I had always understood it, became irrelevant. I knew where I existed in any given day or night—knew my place in time—by the position of the sun and the constellations. That was the difference—I oriented myself within the span of days and nights, rather than orienting time around myself.

Two simple truths became immediately clear: the sacredness of water and the arrogance of wastefulness. I held onto my garbage, and so waste was a liability—an obvious one. There was nowhere to hide it, no one else to blame for it, and the natural next step was to not create it in the first place. My relationship with water was redefined by a series of daily rituals: collecting water, carrying water, conserving the water I had on hand, using only just enough. Bathing was planned for in advance—I filled a three-gallon milk pail from the spring, left it in the sun all day to warm the water, and bathed in the late afternoon while the air still held some heat.

There was a direct relationship between the water and my body.

Walking to the spring; lifting the heavy wooden lid of the spring box; lying on the ground to dip a pail into the clear, wild water; filling two three-gallon milk pails bucketful by bucketful; safely closing up the spring box; lugging the full pails, which together weighed fifty pounds, back to my camp; setting them carefully on the shadow side of the camp trailer. Without the intermediary of a faucet, I was granted an embodied understanding of the gift I was receiving, which made that gift impossible to take for granted, impossible to waste. Every drop mattered. Every drop was sacred. When I'd moved into the cabin, I had learned how little I really needed. On the mountain, I realized it was even less. Having just enough is everything.

After milking Daisy each morning, I filtered her milk into gallon and half-gallon jars, which I carried to the spring box to chill. One morning, I held back a half gallon because I planned to make pancakes with eggs Mike had brought up. After I ate, I picked up a book instead of taking the leftover milk down to the spring box, thinking, *This must be what it's like to live in a mansion—an excursion just to get to the refrigerator.* The nearly full jar sat on my countertop. The cream rose to the top, and the next morning, I spooned some of the cream into my coffee. The cream had lightly cultured to become crème fraîche. Coffee, swirled with a dollop of crème fraîche, was delectable, and I left the jar on my counter. The next day, after another cup of crème fraîche coffee, I made cultured butter from the remaining cultured cream. I packed the butter into a small dish and savored it by the spoonful, every rich mouthful bursting with sweet, multidimensional flavor. And I left the jar of milk on my counter. I ignored it the following day, and the day after that, I lifted the lid and tentatively sniffed it. The milk had a distinctly sour scent. It didn't smell like rotten pasteurized milk; it was more of a sharp tang, like yogurt. I dipped a spoon in the jar and tasted the milk. The flavor was not unpleasant, just very sour. I wondered if perhaps there was enough acid in the jar of sour milk to make a soft cheese. To make quick soft cheese, milk is mixed with white vinegar or lemon juice and gently heated until it separates into curds and whey. I hadn't thought to bring white vinegar to the mountain with me. I poured the sour milk into a pot, set it over low heat, and in twenty minutes, I had curds and whey. Charlie and Chloe got the whey, and I had a delicious soft cheese, which I seasoned with salt, pepper, and thyme.

I stopped putting milk in the spring box after that. My little countertop was filled with jars of sprouts in various stages of growth, and jars of milk and cream in various stages of culturing. Daisy's raw milk transitioned through unique stages without anything added but warmth and time. The early crème fraîche went in my coffee and tea. I made cultured butter when I had the patience, and fresh cheese from large batches of soured milk every few days. Lentil and pea sprouts mixed with fresh cheese and drizzled with olive oil and balsamic became a daily favorite. An ancient, delicious world had opened up to me, purely by accident and experimentation.

I walked everywhere. Frisco often joined me. He tagged along on my trips to the cell phone spot, walking beside me through the sun-dappled woods, his cheek at my shoulder. Warm pine sap perfumed the air, like spiced honey on the breeze. The treetops swayed high above us in gentle sighs. When we reached the clearing, Frisco grazed while I called Mike for our daily check-in or caught up with friends. When Frisco gracefully plopped down to rest, I lounged against his great shoulder, tapping out blog posts and emails while he chewed his cud. I've always been jealous of cats, that they get to curl up in a person's lap, encompassed by the warmth and life of another being so much larger than they, one they trust enough to fall deeply asleep. Curled up against Frisco, I got to be the cat.

Frisco and I explored the mysteries of the mountain without destination or direction other than moving where the land suggested it, following deer trails while serenaded by sage chickens, woodpeckers, and ravens. With Frisco by my side, I found and feasted on wild strawberries and gooseberries, collected feathers to adorn the camp trailer, watched a hundred monarch butterflies dancing upon a hundred purple thistles. We sat in the sun on the sandstone bluff, watching it rain in the valley, wispy curtains of mist drifting our way. We discovered an echo at the edge of the woods. Frisco's moos sailed across space and returned to me, the vibrations of his voice thrumming in my chest.

In the weeks leading up to our move to the mountain, I had driven to the pasture a number of times, alone and with Mike, to see if the truck trail was dry enough for the camp trailer, to fix up the corral for milking, to make sure the move with all my animals would be smooth and easy. I had been startled, but not surprised, by how little evidence of wildlife I saw during my trips through the national forest. RVs lined the shoulders of the roads. Technicolor tents and camper vans were scattered in the pines like mutant mushrooms. The campgrounds were swarming with people, the roads noisy and dusty from a steady flow of ATVs. Even the birds seemed to stay away. After I passed through a few gates and was deep into private land, wildlife emerged in breathtaking abundance. I saw the tracks of coyote families on the paths I wandered—pups and parents traveling together, evidenced by the sizes of the paw prints in the dirt. I stumbled upon the scat of mountain lions and moose, and was granted daily sightings of deer. While picking chokecherries, I noticed torn chokecherry boughs and parallel scratches on the trunks of neighboring trees, signs that a bear had been gathering fruit, too. I saw more species of birds than I could count, more butterflies than I could name, and more rattlesnakes than I would have wished. Coyotes sang nearby each evening, and one night, I heard the long, lone howl of a wolf, unmistakable in the darkness. Mike told me later there had been two sightings of a white wolf about five miles from my camp.

US Forest Service biologists have noted compromised wildlife habitat on public lands, even in areas with leave-no-trace and hike-in-only access. There are people who denounce all animal agriculture as unethical, but one could argue that private ranchland is one of the last sanctuaries for wildlife, free from excessive human encroachment, because a ranching family checking cows periodically, or even my camping in the mountain pasture all summer, has far less impact on wildlife than the influx of thousands of hikers and campers entering wildlife habitat in public wilderness areas over the course of each season. Still, private ownership of land does not guarantee that wildlife and natural ecosystems will be nurtured. Blanket statements are dangerous and never convey the full truth, and conservation is incredibly nuanced.

Many people equate conservation with removing humans and our impacts altogether. But even if an area is fenced off, it's impossible to remove our impact on the landscape. We have drastically altered the lay of this land. They say you could perch on a butte and watch a herd of buffalo thunder by in the valley below for four days straight. That's impossible now, not only because the buffalo were nearly extinguished by European colonizers. We've divided this land into too many pieces for a herd that large to roam freely. We've changed this place. And that has consequences. We cannot say to the natural world, "Here, go back to the way you were—but you have to stay within these certain boundaries to do it, and you must work around the dam upstream, and the highway over yonder, and the shopping mall we built in the middle of historical migratory paths. Now, go forth and be wild!" Left alone, in this fractured state, the natural world can become dangerously out of balance. When forests are neglected, wildfires are ever more potent. When a predator population outgrows its prey, or vice versa, they wreak havoc on their environment until they eventually starve themselves out. When the balance is off in one area, it throws the balance off everywhere, because everything in the natural world is interconnected—including us.

Humans have been modifying and interacting with their environments for millennia. And while central to American lore is the idea that this land was a bounty of natural resources untouched by humans before Columbus arrived, the truth is that Native Americans were here for tens of thousands of years, interacting with their environments on a grand scale, with managed fire being one of their most powerful tools, and the bounty existed, to a significant degree, *because* of their influence. As evidenced today by surviving Indigenous cultures—who make up 5 percent of the world's population and whose lands, under their stewardship, hold 80 percent of the biodiversity on the planet—the interplay between humans and the environment can nurture and even improve ecosystems rather than destroying them. The heart of the matter is the relationship.

The consolidation of land into the hands of the few is taking place at an alarming rate. In the United States, the amount of land owned by the one hundred families with the largest holdings totals forty-two million acres. And this is a 50 percent increase from 2007. Is the goal to caretake or control? Belonging to a place is entirely different than owning a place. When you belong to a place, you are in a relationship with the land. Being in a relationship requires attention. It entails understanding what that land holds, what it gives, what it needs. Relationship with a place, with an ecosystem, is like any other relationship. Attention grows into familiarity. Familiarity grows into intimacy. Intimacy elicits a reflexive urge to care for, protect, and defend that place, and to protect everything else that belongs to that place. Is it possible for one person to belong to seven hundred thousand acres? To two million acres? How could one person possibly establish intimacy with that vast an area? If one were to spend ten hours a day, seven days a week, walking the land in twenty-foot swaths at the average rate of three miles an hour, it would take twenty-six years to traverse seven hundred thousand acres. Twenty-six years to see it *once*. There's no intimacy in that. There's no opportunity to decipher the messages left by tracks in the snow and mud, to learn the secrets held in the soil, to give attention to the flora and fauna—to notice what arrives, what returns, what leaves—and through that attention, to learn the behaviors and needs of the ecosystem as a whole, along with each part, in order to tend and encourage it.

For comparison, by the same calculations, it would take a week and a half to traverse the eight hundred acres of mountain land where I summered. Which I did. Week by week, month by month, I spent the summer afoot, learning the contours of the meadows and the favored routes of the wind just as I learned every barbwire scar on Mike's forearms. It was just beginning. There was so much more to know. But I felt myself belonging.

Autumn on the mountain arrived in early September. Birds began to disperse, their trills and chatter replaced by the bugling of elk. Overnight temperatures dropped below freezing, though it was still decidedly summer in the valley. Eli crawled under the covers at night, and I could see my breath when I woke. Every leaf and blade of grass glittered with frost until the midmorning sun warmed it away. One morning, as the sun crested the highest ridge, Chloe started barking a frantic alarm. I launched myself out of bed and poked my head out the door of the trailer to see what she was announcing. Two bull elk regally traversed the meadow, their massive antlers edged in pink from the sunrise sky. They paused to return my stare, then continued on through the cows to the trees.

Even midday, the air had an edge that set my cheeks tingling.

Summer was gone and, with it, the acute fire danger. Nightfall arrived much earlier than it had in midsummer, and I met the dark with evening fires built from fallen branches gathered in the woods and hunks of dead sagebrush that burned like they were infused with gasoline. I sat beside the crackling fire watching the sparks sail skyward, joining their star sisters in the darkness above. I had been on the mountain for three months. The time felt as rare and magical as Alice's tumble down the rabbit hole. I felt free, joyful, more relaxed and more enlivened than I had in years. Not working full-time had something to do with it. But I hadn't been idle. The nature of "work" was completely different on the mountain. Washing my clothes and sheets by hand hadn't felt like a chore; it was merely something that needed to be done, and was best done in the sun, and completing the work gave me a sense of calm satisfaction, and the mild cardio was a bonus. I carried water from the spring. I prepared all my meals from scratch every day. I walked two miles round trip to make a phone call. The peace that imbued my bones and my soul was not achieved by avoiding labor; it had more to do with the nature of modern work and modern stress. I had temporarily escaped a certain type of work—not work itself.

I ran away, and I found myself, re-collected myself. I found pieces of myself in the rainbow chert, in the glittering spring water, in the scent of mountain thunderstorms, in the sighs of mule deer, in the eyes of a rattlesnake, in the ruby contours of wild rose hips, in the sap-sticky pinecones that stuck to my hair when I lay on the forest floor. I didn't realize how fractured I'd been until the wildflowers and the colors of sunrise put me back together.

And then it snowed, light flakes falling all day, the sky a soft, even gray from sunup to sundown. The snow melted within days, but it was a warning. If a big storm blew in and made the truck trails impassable, we'd have to travel down on foot. It wasn't even a question of "if," but of "when." I didn't want to leave the mountain. I wasn't ready. But Mother Nature calls the shots. It was my job to pay attention, understand her signs, and follow her lead.

I moved off the mountain in one day. I didn't want to drag it out going back and forth as I had when preparing to move up—it would have been too hard on my heart. It was time to head down, and I just ripped off the Band-Aid. Mike came up the final morning with the horse trailer. I needed to move Daisy and Fiona back to the valley with me since I was milking, and I wanted Frisco with us, too. Sir Baby would trail down a few weeks later with the rest of the cows. We loaded Daisy, Frisco, and Fiona in the trailer and drove them home first. I rode with Mike. It was the first time I'd been in a vehicle in months. We took the back road off the mountain—truck trails the whole way, which made the trip twice as long but avoided the scream of the highway, which I wasn't ready for. I found I wasn't prepared for truck trails, either. At one point, clutching the armrest, I turned to Mike and said, "You're driving *really* fast," and glared at the speedometer. He was going 15 mph.

When we pulled up in front of my cabin, I started crying. Endings are as hard and as fraught as beginnings. Maybe more so. We opened the trailer for Daisy, Frisco, and Fiona and they trotted out, seemed to shrug with nonchalant recognition at being suddenly delivered home, and began to graze. Mike and I unhitched the trailer and headed right back up the mountain, but we took the highway because we still had much to do and no time to waste. I slid onto the floor of the passenger seat under the dashboard to finish crying and to block the speed of travel. By the time we got back to the mountain, my tears were drained, the proverbial Band-Aid fully off. I still didn't want to pack up and go, but I was capable of doing it.

We gathered everything I had brought up and made several checks around the camp to make sure I didn't leave anything that hadn't been there before I arrived. I secured my belongings in the camp trailer for the bouncy trip down, and put Chloe and Eli in my truck. Mike headed down with Charlie in the seat beside him and the camp trailer on his hitch. I wasn't ready to navigate the highway behind the wheel, considering I could barely handle it as a passenger, so I took the back road off the mountain. And then my brakes went out. With a three-thousand-foot descent ahead, I put my truck in four-wheel drive and low gear so the transmission could do what my brakes could not. I rolled down the mountain at about five miles an hour, which happened to be the speed I was comfortable with, anyway. Eli curled up in my lap and Chloe slept on the seat beside me. Taking it slow, down the empty dirt road, felt like the only appropriate way to leave the mountain. When I reached the valley, my magical summer on the mountain would be over, and I was in no hurry to reach the end. It was 10:30 p.m. when I pulled into the driveway. Mike had unhitched the camp trailer next to the cabin, and the animals and I went directly into the camp trailer and slept the night there. I could stave off the ending for one last night, as long as I didn't go in my cabin.

The next morning, when I walked into the cabin, I saw my reflection in the mirror on the wall and stopped and stared. I hadn't seen a mirror in three months. Occasionally, I'd caught a glimpse of my reflection while walking past my truck's dusty windows, but my appearance was so completely inconsequential on the mountain, I had forgotten to pay attention to what I looked like.

On the mountain, my impact had been obvious and, thus, my significance. But my sense of self-importance receded as my days on the mountain turned into weeks, then months. It wasn't that I felt unimportant, but that I felt no more important than any one of the pine trees that made up the woods. It's hard to maintain a certain threshold of self-importance after interacting so directly with the source of what we depend on for our very lives—the sacredness of spring water, of dry wood for a fire, of juicy gooseberries found and foraged, of light from the sun. By and large, we've left the environments that demonstrate, at every turn, the extraordinary significance of all that exists outside of ourselves—all that makes up the natural world around us, all the elements and beings that support life on this planet in a mind-blowingly intricate dance of cooperation and reciprocity. In the environments we've created for ourselves—home, office, car, store—our needs are met through infrastructures and intermediaries made by man, which make it so easy to place ourselves at the center of focus. Even if that focus is subliminal, this can warp our perception, artificially enhancing our sense of importance as individuals and as a species, leaving all the rest, including our impacts, in the background, a blurry backdrop to our lives.

Ironically, or perhaps not, it was on the mountain, unhitched from mirrors, wandering further afield from my ego with each passing day, that I, for the first time, recognized my body as sacred. I recognized my body as sacred for all it allowed me to do—namely, the simple yet wildly improbable act of existing. I recognized my body as a fractal element of the Earth herself, the same notes on a different scale. I recognized that there was really no separation, no division between us, save one crucial, irrefutable exception: the Earth can survive without me, without humans whatsoever; we cannot survive without her.

You are personally responsible
for becoming more ethical
than the society you grew up in.

—ELIEZER YUDKOWSKY

CHAPTER 5

Nature as a Model

never knew how special a blade of grass really was, before I got Daisy. She nibbled tender tendrils when the grass shot up the first warm week of spring. She wrapped her speckled tongue around the bouquets of rogue grass I pulled from my garden pathways. She lounged upon the cool cushion of pasture grass on hot midsummer days, chewing her cud. She figured out how to open the gate to Mike's fenced yard and mowed the lawn with her mouth. Watching Daisy gobble great mouthfuls, munching in chlorophyllic ecstasy, made me look at the generic, green growth with new eyes. Turns out, it's not so generic, after all.

Grasses grow in nearly every climate and terrain around the world. Native grasses are hardy, able to survive flooding and drought, and flourish in all manner of soils. They grow where crops cannot. While the leaves of grass feed numerous species, much of the magic of grass lies beneath the surface of the earth. Native grasses are perennials, which means they grow back year after year without having to be replanted. This longevity allows perennial grasses to expand their root systems over the course of years, decades, centuries. Roots of established perennial grasses can measure over ten feet in length. Their massive root networks serve to anchor the plant, and also anchor the soil, protecting it from erosion. This subterranean tangle of roots support mycorrhizae, fungi dancing in the dark, which have a symbiotic relationship with the root systems of many plants. Mycorrhizae deliver nutrients from the soil to plants in exchange for carbon, which plants draw out of the atmosphere in the form of carbon dioxide through photosynthesis. Mycorrhizae use that carbon to produce a substance called glomalin, which coats and protects the intricate filaments of mycorrhizae called hyphae. Glomalin improves and stabilizes the structure of soil, making it more porous, allowing water to easily infiltrate the ground. This not only helps soils retain moisture to nourish the plants, but also prevents flooding during heavy rains or sudden snowmelt since carbon-rich soil can soak up water like a sponge.

Mycorrhizae are the unsung heroes of sequestering carbon in soil. In our quest to head off increased global warming, sequestering carbon—removing excess carbon from the atmosphere and stabilizing it elsewhere—is arguably as important as reducing fossil fuel emissions. By cooperating with the root systems of plants and producing glomalin, mycorrhizae secure harmful atmospheric carbon in the soil as stable organic matter. If undisturbed, that carbon can remain sequestered in the soil for thousands of years.

Cattle are demonized for destroying the environment through overgrazing and greenhouse gas emissions. But the truth is far more nuanced, and it's imperative we understand the truth. Among all the food we eat, the carbon footprint of beef from concentrated animal feeding operations (CAFOs) is second only to that of lamb. The familiar statistics regarding the harmful environmental impacts of beef are based on beef from CAFOs, which currently makes up the vast majority of beef available on the market. These statistics are true, but cattle are not the culprit. The fault lies in how those cattle are fed and managed when they reach CAFOs. The method is the madness.

Cattle themselves are not inherently bad for the earth or the environment. On the contrary: by keeping cattle—and sheep, for that matter—exclusively on pasture and managing their grazing thoughtfully, these animals can be a formidable force in sequestering carbon and protecting us all from climate change. Beef from responsibly grazed cattle has been proven to have a net-negative carbon impact. That means cattle help sequester more carbon through grazing than the combined emissions of methane from their burps and the fossil fuels used in ranch work, butchery, and delivery. The stewards at White Oak Pastures in Georgia have determined that, pound for pound, their pastured beef has a carbon footprint 111 percent lower than CAFO beef and sequesters nearly twice the amount of carbon than that released in the production of soy. Bovines are one of the most promising and effective allies we have to help heal the devastating mistakes we've made since the dawn of the industrial age.

Mike and I leased a pasture that had previously been used solely for hay crops. Grazing animals were never kept on this particular pasture. The landowner grew a mix of grass and alfalfa, which he treated with synthetic fertilizer and pesticides, and when it was tall enough, he cut and baled it. It was an adequate pasture. When we took over the lease, we asked that the fertilizer and pesticide regimen no longer be used, and we trailed some of our cattle to the pasture to graze it for the summer and fall. When it snowed, we fed hay in the pasture, unrolling giant round bales in a stripe down the length of the pasture. Each day, we unrolled hay just to the left of where we'd fed the previous day, eventually crossing the pasture lengthwise before we brought the cows home to calve. The next spring, the pasture grew in thicker than it had the year before, the grass and alfalfa a deep emerald green. The year after that, the soil was nearly impossible to see when we walked the pasture. Grass had filled in the bare patches and reached our thighs, and the earth was spongy beneath our feet. This improvement in the land—in the soil and the amount of water it held, in the flourishing growth of the grass—didn't take place because Mike and I are particularly special. It happened naturally, because that's how nature works when it's allowed to.

Grazing animals and plants and the universe of microscopic life beneath the surface of the soil coexist in harmony, give unto each other, receive from one another, and in this relationship, each part becomes stronger and healthier. The whole becomes stronger and healthier. As they grazed, our cattle dispersed their natural fertilizer evenly across the pasture in the form of their manure. They trampled old plant matter into mulch, which protected the surface of the soil, slowing the evaporation of moisture, and composted over the seasons, further fortifying the soil. The surface area of Daisy's four hooves—where her mass meets the ground—is nearly identical to the surface area of my two feet, and she weighs ten times as much as I do. All that weight, distributed onto four cloven hooves, works to aerate the soil

and tamp grass seeds into the earth. The very act of grazing stimulated the grass to grow, just as mowing a lawn will do, which made the pasture more productive and resistant to weeds.

After three years, the pasture was so lush, the landowner kicked us off because he wanted the hay crop. We watched from afar as the pasture declined in the years to come, after cattle were removed from that patch of land. Without grazing animals depositing dynamic, bioavailable fertilizer and controlling weeds by eating them before they went to seed; without their hooves aerating the soil; without the light mulch of hay left over from feeding cattle across the pasture in the winter, the vitality of the living organism of the pasture itself deteriorated. Without cattle on the land, returning nutrients to the soil with their manure, the landowner had to resume his use of synthetic fertilizer and pesticides, which degraded the soil ecology, and both the vibrancy and the yield of the pasture steadily declined.

We all know that animals eat plants. But what many people don't realize—I didn't, myself, until I was in my thirties—is that plants eat animals. Plants need more than sunlight and water—they also need food, nutrients to survive and thrive, and animals provide these nutrients to plants with their manure, their blood, their bones. Cover crops, compost, and worm castings also supply nutrients to the soil, but in modern agriculture, the most widely used method for providing nutrients to plants is through synthetic fertilizers.

Creating synthetic fertilizers begins with mining for minerals and drilling for natural gas. The measures involved in extracting and processing phosphate and natural gas make synthetic fertilizers one of the biggest contributors to the carbon footprint of food. Industrially farmed monocrops—enormous swaths of land planted in a single crop—depend on synthetic fertilizers to return nutrients to the soil to feed the plants. In the United States, the vast majority of corn, soybeans, nut trees, fruit trees, and grains are farmed as monocrops.

Just as there is a chain of interlinking actions and reactions between grazing animals, grasses, mycorrhizae, soil, and water, the use of synthetic fertilizers sets off an equally impactful, yet significantly different, chain of events. Synthetic fertilizers flood the soil with nitrogen, phosphorous, and potassium—the macronutrients plants need to grow. Plants absorb these nutrients directly from their roots and abandon their relationship with mycorrhizae. No longer receiving carbon from the plants in exchange for nutrients from the soil, the mycorrhizae suffer, cease weaving their glomalin-wrapped webs of hyphae. Without glomalin aggregating the soil, the soil becomes compacted, losing its capacity to absorb water. This leaves the land vulnerable to flooding, and more irrigation from rivers and aquifers is required to water the plants. Irrigation from aquifers deposits mineral salts into the soil; when they accumulate, the soil becomes too saline to grow crops. And without their relationship with mycorrhizae, plants do not receive micronutrients not present in synthetic fertilizers. Nutrients not available to the plant means nutrients not present on our plates.

Annual crops, such as corn and soy, must be planted every year. Annual monocrops require herbicides to kill competitive plants when the crops are young, and enormous swaths of land planted with a single crop are particularly vulnerable to damage from insects, requiring insecticides. Pesticides—herbicides and insecticides—are made from petroleum. And while perennial pastures are covered in grasses year-round, monocropped fields leave the earth bare for months out of the year. Even when crops are in the ground, naked earth surrounds each plant in a monocropped field, each tree in a monocropped orchard. It takes effort, be it mechanical or chemical, to keep plants from growing in bare earth. The soil wants to hold and be held by roots.

Without these roots, wind and water carry away the topsoil. Topsoil, essential for growing crops, is being lost at an alarming rate from tilling and erosion. The topsoil beneath the prairie grasslands that once covered the Midwest was six feet deep when European colonizers took control of the land. Scientists estimate that, at the current rate of erosion, our topsoil could be completely gone in sixty years. Laden with synthetic chemicals from fertilizers and pesticides, the runoff from monocropped fields pollutes waterways and has formed a dead zone larger than Rhode Island at the mouth of the Mississippi River. Rates of cancer soar along its banks. Honeybees and pollinators die en masse from exposure to pesticides, while the insects that threaten crops develop resistance to these poisons. Phosphate mines in Florida are surrounded by man-made mountains of radioactive waste material—a by-product of the manufacture of phosphorous fertilizer—for there is no way to dispose of it safely. Twice, sinkholes have funneled this waste into the aquifer that provides drinking water to neighboring communities. Fracking for natural gas, the base of nitrogen fertilizer, has been shown to pollute groundwater supplies and leave surrounding areas prone to earthquakes. Debates rage over drilling in the Arctic and in national monuments. Pipelines expand. The earth heats up. Biotech companies scramble to develop and patent drought-tolerant crop varieties as climate change threatens current yields.

When I rode my Vespa across the country, I took secondary highways across Iowa—long, flat, perfectly straight roads that cut through monocrops of corn extending to every horizon. That leg of my ride was distinct in its monotony. I remember how eerie it felt to be surrounded by plants—stalks of corn in orderly rows like soldiers—yet nothing felt alive. As I held a steady line between the cornfields, I noticed an absence of birds, no cottontails scampering across the road ahead of me, no butterflies flirting in the air around me. It was the only stretch of my six-thousand-mile ride where I didn't see roadkill. I never enjoyed riding past animals that had been hit on the road, and sent out a telepathic prayer whenever I did. But the total absence of roadkill brought a greater sorrow—it was visceral proof that no wildlife survived in this land of monocrops, a graveyard to everything but corn.

Can we really use the term agriculture when it comes to monocrops? "Agribusiness," certainly. "Agrinsanity"? I think so. But it's not agri*culture*. A culture is defined as a symbiotic living system, derived from the Latin *colere*, which means to tend to the earth and is linked to the concept of nurturing, of actively fostering growth. Monocrops are not ecosystems. Very little but the intended crop survives on monocropped land. Incapable of supporting a multitude of healthy relationships—from the microorganisms in the soil to the atmosphere of the planet—monocrops depend on extraction, require destruction in order to exist.

Monocropped corn and soy are cheap to buy. They are easy to store and ship. And so, industrial agriculture takes animals off the land and confines them in CAFOs where they are fed that corn and soy. Early in my cross-country Vespa trip, I rode past a cattle CAFO in the Central Valley of California. An unbearable stench nearly knocked me off my scooter as I approached the tragic wasteland. I accelerated as fast as I dared and took short, shallow breaths as I zoomed past; the olfactory assault a marker, escapable only by distance, of the misguided mismanagement inherent to this system. Pastures of grazing cattle don't stink. Manure is dispersed by the cattle and immediately begins to decompose with the help of beetles, earthworms, and microorganisms in the soil. But instead of being integrated into a holistic ecosystem as a beneficial resource that fertilizes the soil, the manure of animals in CAFOs piles up or is held in gigantic manure lagoons, off-gassing methane as it sits. In the United States, methane emissions from "manure management" have risen nearly 70 percent since 1990, while emissions from enteric methane (bovine burps) have increased by just 8 percent. Grazing and foraging animals don't create manure management problems— CAFOs do. CAFO manure repositories emit other toxic pollutants— ammonia, hydrogen sulfide, particulates carried on the breeze—which directly impact neighboring communities. Reports commissioned by the Centers for Disease Control and Prevention show "repeated exposure to CAFO emissions can increase the likelihood of respiratory diseases," particularly in children.

Enteric methane—the methane emissions for which cattle are responsible—is the result of grasses biodegrading through a microbial process in the animal's rumen. If the same grass were to remain un-eaten, a similar amount of methane would be released when that grass decomposed in anaerobic conditions without the cow, as in wetlands. Enteric methane from cattle—from all ruminants—is simply part of the natural carbon cycle. And unlike carbon dioxide emissions, which accumulate in the atmosphere where they can remain for thousands of years, atmospheric methane only lasts for twelve years. If the cattle population remains stable over decades or centuries, the amount of methane in the atmosphere from enteric emissions also remains stable—to put it very simply, as new methane is added, old methane is dispersing. Enteric methane emissions in the United States have re-mained notably steady for the past thirty years, corresponding with the stability of the cattle population. I'd venture to guess our enteric methane emissions are not much higher than they were for this region five hundred years ago. Before European settlers massacred the buffalo population to near extinction and wiped out much of the elk, it's estimated that as many as seventy million large ruminants roamed what is now the continental United States. Today, our cattle population is around ninety million, and nearly half that number are calves. Meanwhile, atmospheric methane levels have risen sharply since 2008. Blaming climate change on enteric methane emissions from cattle is not only incorrect and misguided, it is dangerous. Doing so deflects attention from the actual causes of increased atmospheric methane, and shifts the blame from where it truly lies.

In 2008, journalist Michael Pollan warned that our conventional food system couldn't last. He predicted the price of fossil fuels would rise enough to render our current model of industrialized agriculture economically unfeasible. Turns out, the disaster we're facing isn't the price of fuel, but the effects of unhindered extraction and consumption of fossil fuels. While I understand a personal choice to not eat meat, I don't understand those who disbelieve in ethical animal agriculture. It is in our collective best interests to shatter the myth that animal agriculture is inherently bad for the planet and growing crops is inherently good. One cannot oppose animal agriculture and also oppose pipelines, fracking, and the expansion of fossil fuel extraction. The soil must be replenished with nutrients in order for plants to grow, and this can come through keeping animals on the land or through the use of synthetic fertilizers. One or the other is required to feed plants on a large scale. The most expedient way to fertilize great expanses of land while decreasing our dependence on fossil fuels is to integrate animals onto the land, to allow them to fertilize it themselves, and to recognize that we need animals to help us if we are to minimize the carbon footprint of our food.

Just as cattle are not inherently bad for the environment, corn and soybeans and almond trees are not inherently harmful, either. It's the way in which they are grown that determines their impact. A mycelial network of stewards across the country are farming and ranching—often, integrating both—using sustainable and regenerative techniques, their methods as individual as each plot of land, each microclimate, each budget. What they all have in common is a commitment to nurturing the health and longevity of the Earth and her resources, recognizing their roles as partners with the land and water, the plants and animals, the soil and mycorrhizae, as they provide food for their communities.

When we think about restructuring our agricultural systems, as we must, we can look to nature as a model. Nature is an exquisite demonstration of cooperation at the micro and macro levels—incalculable species working together for the health and prosperity of the whole. In the summer, I watch the sun rise over the pond below Mike's house. The pond is home to frogs, who serenade us at night. A blue heron arrives each morning and wades in the shallows at the edge of the pond. Beak flashing in the sun, the heron strikes through the surface of the water, spearing frogs for breakfast. I love the frogs, and I love the heron. If you had to choose, would you choose the frog's life, or the heron's? Watching nature shows as a kid, I never could pick a side—as soon as I began rooting, in my mind and heart, for the antelope, I knew I was cursing the lion. And so I'd switch my allegiance to the lion and, in the next moment, realizing this meant death for the antelope, switch back. I wanted both to win, I wanted both to survive. And the thing that seems challenging for many of us to realize and accept is that when nature is in balance, both the antelope and the lion *do* win. They survive, overall, in perpetuity, precisely because individuals do not. Nature isn't personal.

A neighbor's cow died in a remote area of the BLM where I hike. Her body was left where it lay, and I watched her become part of the land. Coyotes and golden eagles ate her flesh. Her bones bleached out from sun and time. Two years after she died, a luminous sweep of dense, lush grass stopped me in my tracks. The circle of emerald filled a radius of several feet from where her body had lain. Her skull was barely visible in the tall, thick grass. The vitality of the decadent growth, nature's shrine to the cow, glowed in startling contrast to the patchy, sparsely vegetated high-desert landscape of the BLM. Plants want our bodies. They want our blood and bones. And this is far from gruesome—it's the truest beauty, the circle and cycle of life. Perhaps it's the very meaning of life: to live in a way that enhances lives beyond our own, and to contribute to life even after we die.

If the human experiment is to succeed, we must watch and learn. Nature gives us a master class: billions of years of practical experience to our ten thousand years of agriculture and seventy years of industrial agriculture. Nature shows us that strength is found through cooperation. That success is achieved through a network of supporting relationships. Nature shows us that death is unavoidable, but that death can, and should, nourish life. Nature shows us that nothing exists in a vacuum, that every action has a reaction. And that every single one of us—a fraction of a billionth of the population—has power and our actions matter, our choices matter.

CHAPTER 6

Flint and Pyrite,
Stone and Steel

When I bought Mike's calves to keep them from going to CAFOs, I never intended to keep them forever. I never intended to keep them from transitioning to food. Sourcing humanely raised, pastured meat had been important to me for years. In San Francisco, I bought grass-finished beef from a specialty grocer and budgeted for the expense. When I moved to Wyoming and found the local stores only carried conventional beef from CAFOs, an unfortunate irony in a state with more cattle than people, I worked in trade for elk—wild game being the original organic, free-range meat. After two years of helping Mike with his cows, I felt an overwhelming responsibility not only to prevent calves from going to CAFOs, but to offer other people the caliber of beef that I had been looking for—exceptionally tasty, phenomenally nutritious, raised with respect and in harmony with the environment. If the status quo of factory-farmed meat was ever going to change, alternatives had to be available, and I was in a position to offer one.

I leased pasture alongside Mike for my steers to roam and graze, and I bought hay for them to eat during the winter. My steers continued to live in their family herd, peacefully grazing meadows across the valley and up the mountain as they grew, nourishing the land that nourished them. In the spring of their third year, I shared this venture with my online readership and opened preorders for my pastured, grass-finished beef, raised with love and sold in bulk. I offered half and quarter shares of a whole steer, which could be claimed with a deposit while the steers still grazed on pasture, months in advance of delivery. The balance was determined by the final weight of each steer. Delivery was somewhat unconventional—I was only delivering to a set route of ten cities, in which customers would meet the delivery truck at a particular date, time, and place to pick up their bounty of beef. It was a completely different paradigm than shopping at the grocery store: pay in advance, buy in bulk, meet the truck, receive every part of the animal instead of selecting certain cuts. For customers, the entire process required more effort, more planning, more flexibility, more creativity, more money up-front. Before I launched my beef business, everyone in my life thought I was crazy, that it would never work, even Mike.

I marketed my beef exclusively through my blogs and email list. I had an established relationship with my readers and they knew my philosophy and practices, knew how much I loved my cattle. I believed that among my readership, there were many who also cared deeply about sourcing humanely raised, healthy, pastured meat; who would be eager to buy meat directly from someone they knew, if only virtually; who would find value in tracing their food directly to the cattle and land they'd become familiar with through my writing and photographs. I held a quiet conviction that my future customers and I were already linked by shared values and the internet; and I didn't see my business as a risk or a fantasy, but simply as a step in the right direction that we could take together. I started the business I'd wished for, as a customer, nearly a decade earlier when I lived in San Francisco. The response was phenomenal and I sold out in six weeks.

The trust I had established with my readership over the course of several years was essential to the success of my business. Cooperation cannot be achieved without trust. My customers trusted me enough to spend hundreds to thousands of dollars on a product they had never tasted. They trusted me enough to pay months in advance. They trusted me enough to purchase meat in an unconventional way—ordering a year's supply at once and meeting my delivery truck to get it. The amount of trust my customers had in me was humbling and exciting. We've been conditioned to not trust each other. We've been conditioned to trust bright, shiny labels and bright, shiny aisles. Perhaps that's by design—this arrangement is very profitable for those who benefit from it. But does it benefit us? Mutual trust among a group of individuals is powerful, and exponentially expands the realm of what we can achieve.

After months of grazing the lush grass of spring and summer, the time came for my steers to transition from beautiful beasts to nourishing meat. Since I had known from the beginning that my steers' last day would entail a ride in the trailer, Mike and I had occasionally hauled them between pastures instead of trailing on foot so the steers could get accustomed to the trailer, ensuring the ride on their final day would not be a new or stressful experience for them. Trailering the steers to pasture was always a long, hard day for Mike and me, because it took many trips, but it was quick and easy for the animals and they lost all apprehension of the horse trailer. When we gathered them from their summer pasture, they calmly ambled into the waiting trailer. Mike and I drove them to the abattoir in small groups over the course of two weeks.

We drove slowly, on back roads through undeveloped land, so the ride was gentle on the steers. I was glad to go slowly. I wanted to pause time. The road was flanked by rolling hills and red sandstone buttes, and I stared out the window for most of the trip, watching the light transform the red dirt from a dusky rose to electric tangerine as the sun grew bright. Cattle grazed the open range, their calves scampering across the road ahead of us. It was a quiet drive, with no noise from traffic, for we passed only two or three other vehicles during the forty-minute trip.

The small, family-owned slaughterhouse was a squat, square building with a peaked roof and a corral for the animals about to transition. It looked like a quaint, country farmhouse, surrounded by a tidy lawn and bordered on all sides by large pastures. Its interior was gleaming white and stainless steel. Mike backed up to the corral and I jumped out to open the trailer. The steers trotted out and sauntered the full length of the corral, taking in their new surroundings with exploratory sniffs and glances, then gathered together, brothers. The head butcher came out, looked over my steers, then looked into my eyes and said, "I can see these animals are loved." They stood serenely in the corral, chewing their cud—a sign they were relaxed, for cattle don't chew their cud when they're stressed or anxious. I stood alongside them with my arms dangling over the corral railing. They gazed at me with their molasses eyes, glanced at the mules in the meadow across the road. The morning was quiet but for the birds in the surrounding trees, the occasional car, the rambling conversations of Mike and the head butcher at the other end of the corral.

I was in awe of the gorgeous bodies of my steers, soft mountains of muscle and fat, their shiny, black coats slick and taut, gleaming with health. Their slow, deep breaths tempered the rate of my heartbeat. A blue heron glided over us in a great circle. Even in the corral of the slaughterhouse, my steers showed no signs of fear. They had no previous experience of stress or abuse that would cause them to feel fear now. The other people at the abattoir didn't make them nervous, even though they were strangers, because my steers had never known mistreatment at the hands of humans.

I looked at each steer and, holding eye contact, thanked them individually in a silent prayer of gratitude. I pictured them as calves,

calves whose births I'd watched and sometimes midwifed. I pictured them as yearlings, young and free, trotting through wildflowers on the way to the mountain pasture. I saw them standing before me for the last time, the culmination of their physical perfection, strong and mighty and calm. I told them how grateful I was that they were providing me with food for the entire year. I thanked them on behalf of my customers, whose families would be nourished for the year. I have repeated this silent ritual of reverence every year, for every steer. It's an elegy—for, in the words of poet David Whyte, an elegy "is always a conversation between grief and celebration." It's my way of saying grace, of acknowledging that my food blesses me, that my steers bless me with the gift of life.

On our way home, Mike drove slowly, even though the trailer was empty. I closed my eyes and silent tears traveled down my face. I've been providing humanely raised beef to my customers for nearly a decade and I still cry when my steers transition to food. Not from guilt, not exactly from sadness. I cry from the sheer intensity of being so closely involved in the circle of life and death, and death for life. We're all involved in this—every time we eat and drink, no matter what we eat and drink—but often from such a distance we don't feel it. That distance, that disconnect, is so much easier than feeling the enormity of life and death with every bite. But no one is outside it. Stores make it so easy to disconnect from the process of how our food came to be, but the process is potent. It demands acknowledgment and responsibility. It leaves no doubt that waste is disrespectful. In addition to the meat, every organ and every bone from my steers was spoken for.

Every hide went to a tannery. Every hoof went to someone's dog. Even an ear and an eyeball were saved to fuel a child's imagination—one customer had mentioned, via email, that her daughter was in the midst of the Harry Potter books and had asked for an ear and an eyeball with which to make spells. Each steer would feed several people for a year. Each life would sustain so many.

I knew the transition of my steers was not in vain, but that knowledge didn't make it easier. As Mike says, "If you have 'em, it hurts, and if it doesn't hurt, you shouldn't have 'em." There's a common misconception that ranchers are heartless and view their animals as nothing more than commodities; living, breathing inventory; dollar signs on hooves. This type exists; I've met a few. But I've known far more who are wholly devoted to the animals in their care; gruff, rugged ranchers who shed tears when their calves and lambs are sold and shipped. Ranchers aren't saints, but they devote their lives, even risk their lives, in service to the animals they tend. There are no sick days, there are no weekends. There are no days off in bad weather, and swing shifts are inevitable. One year, I had a broken foot during calving season. I woke nightly at 3:00 a.m., tied a bag over my cast, and checked heifers by the light of a headlamp while traversing the muddy pasture on crutches. No one in my town would find this remarkable. Everyone has stories like this of their own. One February, as the frozen river was beginning to thaw, one of Mike's cows walked out on the ice and fell through. Mike leaped into the freezing water with a chain, looped it around her shoulders, and scrambled back to the riverbank. He hooked the chain to his truck and drove slowly up the bank, hauling the cow carefully out of the river. He built a fire beside her—they both needed it—and she survived. Ranchers will speed directly into the path of raging wildfires to cut fences so cattle and wildlife can run and escape the flames. In doing so, they take the chance that the wind could accelerate, that the fire could overtake them, that their truck could break down in the path of the flames, that they could perish while trying to save their herd. Sometimes, the only way to help an animal who is severely injured is to end its suffering—and by that, its life—as quickly and painlessly as possible. It's not heartlessness that makes a rancher capable of doing this. They would consider it utter selfishness to wait for the vet to arrive with a syringe when they have a gun in their truck. Daily, ranchers prioritize the welfare of their animals over their own comfort, unceremoniously honoring an oath of responsibility bolstered by endurance and a forged intimacy with discomfort.

When I started my business, along with the enthusiasm and support came an onslaught of messages expressing shock and anger. "How can someone who has such love and respect for cattle advocate their slaughter?" they asked. I understood where these people were coming from, to a degree. If Daisy had stayed at the dairy, my beloved Frisco would have been veal. All my steers could be pets. Every animal in a CAFO could be a pet. But they can't all be pets. They can either not exist at all, or they can feed people. And if they feed people, they can live a miserable existence in a CAFO—an unhealthy, unethical system that harms the environment—or they can live peacefully on pastures and prairies, roaming freely in family herds, helping to heal and protect the Earth during their time on it. Since they transition to food, the lives of my steers are short, this is true. But their lives are so completely free—free from stress, free from abuse, free from confinement, free from hunger, free from hardship of any kind. For me, the question wasn't, "How could I . . ." but "How could I *not*." I was in the rare position (currently, just 1 percent of the population of the United States works in agriculture) to keep calves from entering CAFOs while offering an exceedingly healthy and sustainable source of meat to those who chose to eat it. I would have felt extraordinary guilt, and felt like a hypocrite, if I had shied away from this work. I started selling beef precisely because I love cattle.

"Would you eat Daisy?" they asked via email and blog comments. It's hard to read tone in text, but these messages always sounded like a sneer. Nourishment comes in many forms, and Daisy has nourished me with her milk and her presence. She helps me manage the rest of the herd, for they all follow Daisy, and Daisy follows me, and this makes moving cattle to the barn or corral absurdly easy. When I drape my body across hers, my system is flooded with serotonin and oxytocin. I want to spend every second I can with Daisy, and this means if and when she dies of old age, there won't be much to eat. Mike's Grandmother cows lose most of their muscle mass before they die of old age. It's the trade-off for them living out their days—they don't provide food for people (though the wild animals and birds are grateful for this arrangement). So, no, I don't plan to eat Daisy. But perhaps what these messages meant, and certainly the better question, is: could you eat an animal you loved? And my answer is yes, absolutely. How different our environment and our health might be if we could trace all our food to origins of love. How nourished our bodies, spirits, and planet could be if all our food was raised and made with love—grown by people empowered by fair wages and safe working conditions, from land honored by sustainable and regenerative practices, with animals respected and cared for throughout the entire course of their lives. Not all my food has origins of love—though each year, I work to increase that—but all the meat I eat does.

I don't name my steers, but that is more an issue of memory than a defense mechanism. Even without names, I see them as individuals. To the uninitiated, they might appear to be identical black bovines, but each is unique and I can tell them apart by the shapes of their faces, the tufts and whirls of their hair, by the way their coats catch the light. When I sit in their pasture, watching them lazily graze, they will, one by one, slowly approach me and, standing shoulder to shoulder, encircle me. Their dark, sweet faces gaze down at me, calm and curious. They sniff my boots and my body. They lick my hair.

It's not about not getting attached. I love my cattle. I eat my beef. I cry when they die. I sell their bodies because I believe in this work. It's all so incredibly complicated. And it *should* be complicated. The neat, sterile packaging in grocery stores removes the general public from the processes of how that meat came to be; absolves the consumer from any responsibility for what happens to the living beings that provide us meat; and allows CAFOs to proliferate with profit as the only driving force. It's not easy to consciously take a life—though we all, in modern society, take the lives of others through our choices, whether we are conscious of it or not. I do this work, and I eat my beef, because I find it the most ethical choice I can make, and I make that choice from a foundation of love. I'm not filled with shame when I cook and eat a hamburger from the meat of one of the steers I have raised and respected and loved. I'm filled with reverence, with gratitude, with humility. I exist, because of them. I am strong and healthy, because of them.

"Hypocrite murderer!" they wrote. "You can't love animals and eat them, too!!!" I wanted to reply with, "You can't love the Earth and also eat food that was produced using agricultural practices that destroy pollinator populations!!!" or "You can't love your fellow humans and also buy produce and nuts grown by massive farm corporations that mistreat and underpay the marginalized people working in the fields harvesting your celery and pistachios!!!" But I didn't. I kept my fingers corralled, and instead, I pondered the chasm these messages represented and wondered how it might be crossed. If all interpersonal relationships can be simplified to Venn diagrams, why do we choose to focus on the areas where we differ and write each other off, rather than focusing on where we overlap, where we agree, and cooperating from there? I have never understood the divide, or the amount of animosity, that exists between vegans and ethical omnivores and ranchers. If we agree that factory farms and CAFOs are harmful, why are we not working together to change the system? If we agree that protecting the environment and reducing the carbon footprint of food are of utmost concern, why are we not working together to that end? Infighting is the best way to make sure nothing changes. Discord only benefits those who benefit from the status quo. To effectively dismantle entrenched agricultural systems—economic, political, and logistical—we need the commanding power of numbers. But how can we organize together, how can we create change together, if we don't believe in each other, if we don't trust each other?

The harm comes when we succumb to what educator Yaya Erin Rivera Merriman recognizes as "the pervasive spell of duality . . . in that paradigm, someone has to be wrong in order for you to be right. So, there is no way to communicate that isn't a battle, and usually an escalating one." Different isn't wrong. Just as it is within ecological systems, cultural and interpersonal homogeny is a weakness. If identical thoughts filled the minds of all 7.5 billion people on the planet, our species would be defined by its frailty. "The question is not how to eliminate conflict, which is impossible," wrote Riane Eisler in *The Chalice and the Blade*, distilling the work of psychologist Jean Baker Miller. "The question . . . is how to make conflict productive rather than destructive." We regularly use the word "conflict" as a synonym for war—for the most violent aggression possible—but these terms are not interchangeable. Conflict is not inherently violent. It's not inherently problematic. Quite the contrary: conflict is a prerequisite for growth. Without conflict, without dissent, there would be no problem-solving, no innovation. The word "conflict" comes directly from the Latin *conflictus*, past participle of *confligere*, which means "to strike together." Strike together. The image these words conjure for me is not of two people punching each other, but of flint and pyrite, stone and steel, striking together to create a spark. What if we saw conflict as the birthplace of transformation, source of the potential inherent to a carefully tended and nurtured flame? What if we realized that to activate that spark of inspiration, we had to clash, and we entered into the clash with anticipation and trust? What if we saw conflict as an act of cooperation?

A week after my steers transitioned, I stood in a meat locker, surrounded by my hanging sides of beef. It was the first time I'd been in a meat locker, and I hid my apprehension as my exuberant butcher led me to the door, excited to show me how gorgeous the meat was, how the healthy fat gleamed like sunshine. The sides of beef towered above me in the giant walk-in cooler where they were dry aging to further develop tenderness and flavor. I was overwhelmed by the intimacy of standing among the bodies of the steers I had known, bodies that would enter my body. In that moment, the anonymity of all the food I'd eaten during all the years of my life felt wrong, felt weird, felt like such a loss. The fact that my relationship with food, up to that point, had been based in obscurity and unknowability was something to mourn. Eating is intimacy. We deserve to know every-thing about everything we put into our bodies. Our ancestors, and not far removed, were intimately involved with all of their food, or sepa-rated by just a degree. Now, the separation is so vast, for most of us, it's often impossible to calculate.

I suspect part of people's discomfort with my business stemmed from our cultural reluctance to acknowledge death. I've spent a lot of time wondering why the world is set up this way—why something must die in order for something else to live. This four-dimensional existence is defined by separateness. I am separate from the tree, which is separate from the rabbit; yesterday is separate from tomorrow. But perhaps, by consuming a cow or a chicken or a celery stalk, all of which must die to feed us, we are meant to be reminded, daily, at the cellular level, of the interconnectedness of everything. That, in fact, nothing is separate; everything is linked. That we are part of it, that we are made of it. With each bite, we are reminded: what was outside of me becomes me. With each bite, we are reminded that every choice we make touches everything, everywhere, and the choices we make today determine the future.

When I launched my business, I hired a professional truck driver to make the deliveries, and she was an invaluable ally in getting my beef to customers that first year. Though I sold literal tons of beef, the load didn't even begin to fill a semitrailer, and the fuel required to power the semi and chill the cavernous trailer during transit was far too wasteful. It wasn't something I could continue in good conscience. The next few years, I rented smaller refrigerated trucks and made the deliveries myself. I loved the opportunity to move beyond the veil of the internet and meet my customers face to face, sharing meals when schedules permitted, spending nights in their guest bedrooms. I enjoyed the annual road trip, stressful as it was with the pressure of a strenuous schedule and the precious cargo in my care. But renting trucks from a national chain proved frustratingly unreliable, and the issue of wasted space and fuel persisted when I was routinely handed keys to enormous twenty-six-foot trucks after reserving the much smaller fourteen-foot trucks months in advance. Shipping directly to my customers via FedEx or UPS was far too expensive without a discounted commercial account, for which I didn't qualify. After much deliberation, I bought a small, used refrigerated truck with the help of a Kickstarter. Owning a reefer truck brought its own drama, from navigating Department of Transportation permits and paperwork to the mechanical mysteries and malfunctions inherent in an older vehicle. I chalked up the time, the expense, and the stress involved in the annual deliveries to the emotional and financial cost of doing business—not simply doing business, but doing what I believed in. As long as the truck delivered the beef safely to my customers and me safely home, I considered it a net success.

Customers arrived to meet my truck with smiles and snacks and coolers and children and dogs. There was an air of excitement and jubilation at every stop, warm introductions with new customers, hugs and catching up with regulars. When I pulled open the door of the reefer, a cloud of frosty air billowed out. I jumped in and carried boxes to the doorway. Everyone patiently waited their turns, chatting with others in line. No one seemed a stranger at these delivery stops, even though most had never met. The strong and energetic helped carry orders to the cars of the elderly or those wrangling young children. Boxes of beef were stacked in back seats and hatchbacks or loaded into coolers for longer rides home. Some people drove six, eight, ten hours to meet my truck, an annual pilgrimage. "It's Meat Day, one of our favorite days of the year!" they exclaimed, thanking me with an earnestness that took my breath away.

There's a kinship: I feed them, they support me. We take care of each other. It's face to face, handshakes and hugs. My business is an intricate choreography between the rain and the sun and the grass and the cattle; between myself and my butchers and my customers; each part equally essential to the whole.

My customers often remark that they came for the ethics and stayed for the flavor. Properly raised grass-finished beef delivers spectacular flavor, rich and nuanced, flavor that is utterly lacking in CAFO beef. It's a myth that cattle can't gain adequate weight on grass alone and that grass-finished cattle are destined to be smaller than those from CAFOs. It does take longer for cattle to fatten on grass than on corn—cattle in CAFOs reach finishing weight about 30 percent faster than my steers on grass, but my steers each provide about 30 percent more meat than the average steer from a CAFO. The meat is exceptionally tender because the animals have been protected from stress, and the fat contains up to six times the omega-3 fatty acids of grain-fed beef—healthy fat that our brains and bodies need. The flavors of the meat, the fat, the organs, and the bones of animals raised on pasture are composed by the landscape. It's no different from wine— pastured meats can be distinguished by the flavors imparted by region, forage, finishing technique, age, and breed. Terroir is the taste of place. Industrially farmed produce and the meat of animals raised in cages have lost their flavor because they have lost their environment. The flavor of my beef develops through the synthesis of soft earth and moonlight, glistening dew and sweetly scented grasses, leaves and flowers found and favored; through the vitality of the animals and the land.

It can be hard for those living in cities or suburbs to connect with farmers and ranchers, but there are many ways to find and meet the people who grow and raise the food you eat. Farmers markets are perhaps the most accessible gateway. A farmers market is a great place to wander, observe, and strike up conversations with vendors—no one should ever be afraid to ask questions. Small-scale farmers and ranchers are obsessed with their work and we love to talk about it. We want to share what we know, what we do, and why we do it. Community supported agriculture (CSA) shares, classified ads, social media, and bulletin boards or contacts through local indie health-food stores and co-ops offer other avenues. It can be intimidating to talk to a rancher, particularly for those who don't know much about meat beyond what

they like or are looking for—local, organic, humanely raised, or grass-finished. But interest and curiosity are all that is needed to start a conversation. Few ranchers are going to shut the door in the face of someone who opens a conversation with "I want to start buying my meat directly from a rancher, but I'm new to this." And when you find someone great, spread the news. Sharing by word of mouth is perhaps the greatest gift you can give an independent farmer or rancher, and it helps others who may be looking for direct, reliable sources, too. Buying in bulk, directly from farmers and ranchers, is often the least expensive way, on a per-meal basis, to get high-quality, grass-finished, pastured meats and organically grown produce, and the convenience and variety of having a year's supply of meat in a chest freezer at home feels like the very definition of luxury. When a personal connection is made, it's possible to go deeper than the label, and engage in relationships with farmers and ranchers whose practices align with our priorities.

Labels never tell the full story. "All Natural" means nothing more than artificial ingredients were not added to the meat and it was not created in a lab. An "All Natural" label does not mean the animal was not given growth hormones or antibiotics. "Grass-Fed" simply means the animal was fed grass at some point in their lives—it does not guarantee that the animal was not fed grain or confined in a CAFO for the latter part of its life. "Grass-Finished" means the animal ate grass till the end of its life. It was once safe to assume that "Grass-Finished" beef and dairy came from cattle that spent their lives on pasture. But recently, it's become common for less scrupulous producers to feed cattle compressed hay pellets while they are crammed in a CAFO, a technique used to capitalize on the growing popularity of "Grass-Finished" beef and technically falling into that category, though the cattle are not on pasture. A "100% Pastured" label means the animal was never in a CAFO, but it could have been fed corn or soy while roaming a barren pasture. Third-party certifications can designate organic practices or the humane treatment of animals, but these certifications are expensive to obtain and many small farmers and ranchers with commendable ethics and practices are not certified, due to the financial burden of doing so, or to avoid having to raise their prices to cover the additional expense of certification. Many people buy grass-finished beef labeled "Product of the USA," thinking they're making an environmentally sound choice or supporting American ranchers, when in reality, that beef could have been shipped from halfway across the world. About 80 percent of the grass-finished beef sold in the United States is imported, but as long as that meat passes through a US inspection plant, it can legally be labeled a "Product of the USA." My beef is "100% Pastured, Exclusively Grass-Fed and Grass-Finished, Humanely Raised Using Organic Practices in Wyoming, and Free from Grain, Soy, Growth Hormones, Antibiotics, and GMOs." It's hard to fit all that onto an eye-catching label. Labels convey but a fraction of what a farmer or rancher can share when relationships are established, even during the most casual conversations while browsing a farmers market, even from the distance of social media.

Every time we buy food, our choices directly impact the environment. Every time we buy food, we fund the agricultural practices that produced it. Buying directly from farmers and ranchers is, at present, the most reliable way to source healthy, nourishing food that was raised ethically and humanely with sustainable or regenerative practices. It's crystal clear what and who your food dollars are supporting. Building reciprocal relationships with indie farmers and ranchers can take extra legwork. It takes being prepared and willing—eager, even—to spend more money, because doing right by the animals and the environment often takes more time and effort and care and money and space. It can take planning ahead and buying in bulk seasonally, which can be much cheaper, and spending extra time and energy preserving or preparing the harvest. It can take using creativity and cooperation to work around logistical barriers. If buying a quarter share of a whole beef isn't feasible, an alternative might be to split it with a large group of friends and doing so a few times a year, which makes each purchase a more manageable financial investment for all involved, and individual shares require less freezer space. Through this arrangement, everyone in the group avoids contributing to the enormous fossil fuel impacts of beef from CAFOs while reaping the benefits of humanely raised, grass-finished beef.

The most sustainably raised foods are often the healthiest foods and often the most delicious, too. These foods are valuable—full of actual, tangible value—value that is demonstrated by the myriad ways they nourish our bodies and the earth. We have the opportunity to make profound impacts—on our health, on the lives of farmers and ranchers, and on the environment—by choosing to invest in these foods.

Not everyone can afford to make such choices. Buying my beef requires more from customers than just the cost per pound. It requires that people have the space and stability to have a chest freezer to hold a year's supply of meat. It requires the ability to pay a large lump sum up-front. It requires paying more per pound than conventional CAFO meat in a grocery store. It requires access to a car to pick up the delivery. It requires having the time to prepare meals at home from scratch, which can be streamlined with practice but can feel like a burden when we are all so busy and tired. One must be resourced enough to make this choice. It's a question of priorities, but only to a degree. If someone is buying jewelry or new clothes every month, or paying thousands annually for climate control in a four-thousand-square-foot house with three people living in it, but balks at spending more money for sustainably raised food, I'll side-eye them. The idea that food should be the cheapest thing we spend our money on is the dark side of capitalism: the system has been rigged to train us to buy cheap food so we have money to spend on more clothes and knick-knacks than we need—often made by exploitative labor—or make payments on a brand-new car. But choice requires a threshold of privilege, and for many, that privilege does not exist. There are very real financial and logistical barriers facing so many, and it's not a question of priorities, it's a question of survival.

In the United States, we spend, on average, less than 10 percent of our household income on food—the lowest in the world. That statistic doesn't tell the whole story, though. The percentage that Americans spend on food is inversely proportional to income, falling to 7 percent for those with an income of $150,000, and shooting to 34 percent when annual income plummets to $12,000. Nutritious food, grown and raised with humane and sustainable practices, shouldn't be exclusive to certain income brackets, but it is. Accessibility to healthy food is directly tied to socioeconomic status—with the possible exception of rural areas where one might hunt or have chickens or a garden. Food apartheid, defined by activist farmer Leah Penniman, is the human-created system of segregation under which certain communities have little to no access to healthy and affordable food, which disproportionately impacts people of color. Any conversation about food is inextricably linked to conversations about racism; class; income inequality; health; the rising costs of housing, healthcare, and higher education; the stagnation of the minimum wage; burnout work culture and our vanishing free time; mass incarceration; voter suppression; tax policy; and campaign finance. It's a web. It's all interlinked. Each part affects and is affected by the rest. This is as true in our societies and economies as it is in an ecosystem.

Recently, I spent a night near the Oakland Airport. The shoulders of the street in front of the motel where I stayed were crammed with cars in which people lived. I went out to get something to eat, but all I found were gas stations and fast-food joints. Walking back to the motel, I passed a blanket and pillow in an overgrown mat of ivy behind a Burger King. The lights from the mansions of billionaires twinkled across the bay. It's all related. Any conversation about food cannot be unraveled from the ways we treat one another and the ways we treat the planet.

CHAPTER 7

A Reflection of the Culture

My art subsidizes my work in agriculture. This would make me laugh were it not so messed up—that art, notorious for being the least viable way to make a living, is more lucrative than providing nutrient-dense, ethically raised food for people. I bootstrapped my beef business nearly a decade ago and have shown a profit every year, but it's the money I earn from my art that pays my health insurance and my mortgage.

Compared with my art, the overhead for my beef business—and for agriculture in general—is enormous. The risk is even larger. Animals can die. Trucks can break down. The weather is invariably unpredictable and uncontrollable. Fuel prices affect hay prices, as do weather and supply. So many things can go wrong. So much is outside our control. As Mike says, "Having cows is a trail of broken hearts and busted checkbooks."

Ranching without owning land is a constant struggle, for the future is inherently unstable. One landowner we leased from sold his land, so we lost that pasture lease. Another landowner decided to get cows of his own, so we lost that lease. Every year is uncertain, and some years we don't know if we'll make it to the next. We're grateful for every lease, for access to land is often an insurmountable barrier to entry for those who wish to work in agriculture. Land prices have risen exponentially higher than income from agriculture alone can support. If land isn't inherited, or financed through a career outside of agriculture, it can remain out of reach for many. If we lost all our leases, Mike and I would have no choice but to sell most of our cows into the very system from which we have worked so hard to shield them. It's impossible to escape an overhanging sense of dread that the life and the cattle we love so desperately could slip from our grasp.

According to a study published by the University of Iowa in 2014, farmers and ranchers in the United States are dying by suicide at 3.5 times the rate of the general population. Based on anecdotal evidence from agricultural communities and the organizations that support these communities, it appears this number is rising. I'm not suicidal, but I viscerally understand this epidemic. Razor-thin margins and a deep sense of responsibility lead to anxiety, debt, and a feeling of powerlessness against forces beyond one's control. Farmers and ranchers don't have control over seed prices, feed prices, or fuel prices; and most don't have control over the price they are paid for the labor and food they provide. The farm economy is treacherous—net income for farmers has dropped by over 50 percent since 2013—and often those doing the most work are those with the least amount of power. Like the loss of topsoil, the loss of agricultural workers to suicide often goes unnoticed by the society they feed. Like the hypoxic dead zone off the Gulf Coast, it's the devastating result of decades of policies and practices that—however well-intentioned they may have seemed at their inceptions—have prioritized yield over life itself and whose legacy is destruction. Improving access to mental healthcare in rural areas and destigmatizing that care can only do so much when the root causes of such hopelessness and desperation are left unaddressed.

For me, it's not the wild weather or the meager profits or even the hollow anguish of losing an animal that pushes me toward the dangerous edge of despair. It's not the work of raising cattle—it's the business end of selling beef. It's the months of each year that I'm stuck in my office, buried in paperwork, on the phone with bureaucrats, coordinating brand inspections, state and USDA certifications, DOT inspections for the truck and DOT medical exams for me—the flurry of permits and paperwork required to sell and deliver the food I raise, with a fee attached to each. During those months, I'm consumed by stress. During those months—when my steers transition to food and that food must get safely to customers—a mantra repeats on a loop in my mind: *Just get through the next few months, just get through the next few weeks, just get through the next few days.* Just get through it.

Sometimes, I wonder why I continue to pour my energy, sweat, and tears into selling humanely raised, grass-finished beef. I would make more money spending that time and energy virtually anywhere else. My life would be so much simpler if I just stuck to art. But I can never stay on this train of thought for very long. I continue because I deeply believe in this work, because I am appalled by the CAFO system, because I am compelled to contribute, in whatever ways I can, to help orchestrate change. Sometimes, that feels like a hopeless goal, my attempts naive and futile. The number of steers I sell, compared to the number of cattle slaughtered each year in the United States, is statistically invisible—too small to have an impact, too small to matter.

And then I think of my cattle. I walk among my steers. I remind myself that, even if the world and the system never change, my work has a significant impact on the lives of each of my steers. Each one is protected from ending up in a CAFO. Each one spends his days in peace and freedom on pasture. I remind myself that my work has a positive impact on the lives of each of my customers, who nourish their bodies and their families with the healthy, humanely raised beef they buy from me. I remind myself that my work matters at the level of the individual, and this keeps me going—for another week, another month, another year.

These reminders don't change the fact that offering healthy, sustainably raised beef has not been healthy or sustainable for me. When the legal and logistical hurdles of sales and distribution exhaust me and I feel strangled by red tape, I climb in the bathtub, pull the curtain tight to hold in the steam, and I cry. The work it takes to achieve what I believe in sometimes feels impossible, the obstacles insurmountable. A dear customer who has been buying my beef for nearly a decade, in whose arms I have sobbed, said to me, "It's OK if you quit. You've moved the needle. You've made an impact." But when I think about throwing in the towel, the pain is overwhelming, nearly unbearable. If I quit, I would be failing so much more than just myself. If I quit, I fail the animals, I fail the land I nurture, I fail my customers, the mission, the future.

The logistical challenges involved in selling meat are a reason CAFOs flourish. Ranchers spend long days tending the land and their animals; many don't have the time or the resources to be salespeople on top of ranch work. And before the advent of the internet, it was virtually impossible for independent ranchers to market their meat directly to customers, especially those not in close proximity to urban markets. And so, most ranchers drop off their calves and cull cows at the local sale barn and get a check in the mail a few days later.

Four transnational corporations, known as the "Big Four" meatpackers, control the sale and distribution of over 80 percent of the beef raised in the United States. They retain contracts with CAFOs that buy cattle from ranchers, and own enormous slaughterhouses and meatpacking plants which together have the capacity to slaughter close to a hundred thousand cattle a day. The Big Four then sell and distribute the meat. The vast majority of beef available in grocery stores, restaurants, schools, hospitals, and prisons comes from CAFOs and slaughterhouses controlled by the Big Four.

What ranchers gain in convenience through this arrangement, they lose in power. Ranchers who sell through the sale barn—which is to say most ranchers—are not allowed to set the price for the cattle they sell. In fact, they don't know what they will earn for the culmination of a year's worth of work until after the sale is made. Cattle prices are determined by the commodity market. A number of factors—such as supply, weather, international trade, and futures trading—determine commodity prices and thus what ranchers are paid for their cattle. Unless ranchers circumvent this system and sell directly to customers as I have, they have no control over what they are paid. A neighbor once ruefully declared that he's made more money from cattle futures on the stock market than he ever has from his cows. Another old cowboy swears that any year ending in the number 3 is a year ranchers are paid fair cattle prices. And it's worth noting that the Big Four are currently being sued by the Ranchers-Cattlemen Action Legal Fund for allegedly violating antitrust laws and colluding to artificially depress cattle prices over the span of years.

Modern mega-CAFOs (be they cattle, poultry, or pig) depend on two fundamental principles working in tandem. Neither is adequate alone, speaking strictly from an economic perspective. The first: animals gain weight quickly and reliably when fed corn and soy.

This is a function of biology, though not necessarily of health. The second: corn and soy are cheaper to buy than they are to grow. This is not a function of biology, nor of the free market, nor of magic. It's achieved through government subsidies. It's vital that everyone who pays taxes, eats food, or is passionate about the environment understands corn and soy subsidies, for they are the keystone to industrial animal agriculture and the bulk of ag-based pollution. Here are the broad strokes:

Subsidies are taxpayer-funded financial support, which are granted across numerous sectors, agriculture being one. Crop subsidies were introduced during the Great Depression as a way to support farmers and stabilize both the food supply and prices. The government bought crops that were easily stored directly from farmers during harvest season, paying them a fair price during an incredibly vulnerable period. The government then introduced portions of those stores to the open market throughout the year, which kept the available supply consistent and the market price stable.

It's been nearly a century since the Great Depression, and the structure of crop subsidies has shifted dramatically. The government moved away from buying subsidized crops directly from farmers to, instead, paying farmers when the market price dropped below a certain threshold to make up the difference. Today, corn and soy subsidies function more like subsidized insurance. Farmers can choose between risk coverage and price loss coverage—risk coverage pays farmers if crops are damaged or lost to pests and weather, while price loss coverage pays farmers when average market prices fall below a predetermined reference price. USDA data from 2018 show that over 90 percent of corn and soybean farmers opted for risk coverage over price loss coverage, likely due to the slim margin between market prices and reference prices for those crops.

Modern subsidy models do nothing to correct oversupply. Globally, we're already producing 1.5 times the food supply we need to feed the world—hunger and starvation are the result of politics, economics, and ethics, not modern agriculture. We actually don't need to farm *more* to feed our population; we need to farm *differently* in order to protect it. Oversupply directly contributes to the financial instability of farmers and the mistreatment of land, while creating a self-perpetuating cycle. As supply increases, prices drop. Lower prices drive farmers to produce more—by any means necessary—particularly if they've opted for risk coverage over price loss coverage. More production means more acreage in monocrops, fewer rest periods to restore the soil, increased use of synthetic fertilizers, more pollution from runoff, more topsoil loss, more ecosystems destroyed. Increased production leads to an even greater supply, which leads to even lower prices. On paper, the farmer is receiving the subsidy. But the parties who are truly benefiting from this system are those buying the oversupply of corn and soy for a fraction of what they cost to raise.

I n 2017, over 180 million acres of cropland in the United States were planted to corn and soy—over 90 million acres of each. It's hard to visualize 180 million acres. To put it in context, in 2017, all the fruits, vegetables, and nuts grown in the United States—sold fresh, frozen, and processed into french fries, tomato sauce, almond milk, and so much more—was done on less than 10 million acres, or about 5 percent of the land used for corn and soybeans.

That's a lot of corn and soy. So what's it all used for?

Depending on the year, 10 to 20 percent of the domestic corn crop is exported. About 40 percent goes to ethanol. About 36 percent of the corn crop, plus spent distillers grains from ethanol production, goes to animal feed, with the vast majority of it supplying CAFOs. Between 4 and 14 percent of the total field corn crop goes to human food products, like candy and soda pop, and industrial use, like packaging. (Sweet corn, which is sold as corn on the cob, canned corn, and frozen corn, is an entirely different crop than field corn, and the sweet corn harvest is included in the acreage used to grow our fruits and vegetables.) Of the domestic soy crop, about half has historically been exported. The primary destination was China, until trade relations became so strained under President Donald Trump that China chose to shop elsewhere. The half we keep is processed into oil and meal. Soybeans are 20 percent oil, and the oil is used in processed foods, for biodiesel, and industrially, in products like adhesives, ink, and foam. The other 80 percent becomes soybean meal, and 97 percent of our soybean meal goes to animal feed. The remaining 3 percent of soybean meal goes to processed human food like mock meats, tofu, and soy milk.

Over 180 million acres of farmland in corn and soy, and most of it is used for exports, ethanol, and confined animals. Exports leave both farmers and taxpayers vulnerable to political instabilities, as demonstrated by the $28 billion bailout from the Trump administration that, as Reuters reported in 2019, went to farmers to "shield them from repercussions of trade disputes." Ethanol, often touted as a "greener" fuel for our cars, is the end result of a process that begins with extracting crude oil from the earth. That oil is converted to diesel to power the tractors, combines, and aircraft that till, seed, spray, and harvest vast expanses of corn, which, as a monocrop, is dependent on synthetic fertilizers and pesticides made from natural gas and petroleum. Once harvested, that corn must undergo a fossil-fuel-driven process, often in coal-powered distilleries, to convert it to ethanol. As the old cowboys say, "You can put your boots in the oven, but that doesn't make 'em biscuits." And animals—beef cattle, dairy cattle, sheep, pigs, and poultry—are designed by nature to forage and graze, requiring little to no fossil fuels for their care and feeding, all while fertilizing the land naturally, helping to sequester excess carbon from the atmosphere, and assisting in the protection and restoration of diverse ecosystems. Instead, these animals are confined to the pens and buildings of CAFOs where they are fed trucked-in corn and soy meal, which is quickly transformed into mountains and lagoons of off-gassing manure.

And we're subsidizing this destructive circus. Corn and soy subsidies, and the overproduction that current models encourage, enable CAFOs to not just thrive but dominate. It's Business 101—the lower the expenses, the bigger the profit margin. Meanwhile, the rest of us foot the bill from every angle: paying for the subsidies, paying for the product (whether it's ethanol at the gas station or CAFO meat), and paying for the environmental consequences for which this system is responsible.

A few years ago, Mike and I were looking at a new pasture lease, walking the land, checking fences, noting the ratio of grasses to invasive weeds. "I saw a bunch of fox kits when I was here yesterday," Mike said, and pointed to a little hill beneath the towering cotton-woods that bordered the edge of the pasture. "Their den is over there, if you want to see if they're out playing." I walked in the direction he pointed but continued past the little hill that housed the fox den, through the trees, to the river. I sat cross-legged in a nest of tall reeds at the water's edge. Their stalks towered above me, hiding me completely. The river swept past, just inches from my feet. I watched the birds gliding in acrobatic arcs above the sparkling surface of the water, deer grazing the wild grasses growing in exuberant tufts on the opposite bank. The sounds of the house and the highway disappeared, replaced by the harmony of flowing water, rustling reeds, chattering birds. Overwhelmed by the lively beauty of this secluded spot, I thought to myself, *Agriculture is the worst thing to have happened to this planet.* This thought shocked me the instant it materialized in my mind. A significant portion of my sense of self is enmeshed in agriculture, so what did this mean about me? What did this mean, period?

As the river tumbled by, the voice of the water unlocked corners of my mind and I thought about it further.

Food is connected to health, to greed, to fear, to insecurity, to economics, to education, to ideas of power, to expressions of love, to the way we connect with each other. Food is our connection to life—of course it is connected to everything else, every aspect of our societies and our psyches.

Agriculture has led to fences, to hoarding resources, to separation. To fights over water and space, to the disenfranchisement and destruction of other people and other species in the name of personal gain. But these are not symptoms of agriculture itself, of planting a seed and watching it grow; of gathering smooth, warm eggs. These are symptoms of fear. These are proof of an overwhelming sense of powerlessness, imagined protections against all that could go wrong. These are declarations of insecurity, for those who feel secure have no need to hoard.

Humans have always had an impact on their environments, as all living things and all elements do. Other cultures, largely Indigenous, have and continue to interact with their environments in ways that support their own well-being as well as that of the surrounding natural world, thereby strengthening the whole. By nurturing the environment that nurtures them, these cultures demonstrate that the needs of humans and the needs of the environment need not be at odds. But taking without giving back in return is not a relationship—it is entitlement, and is, by definition, unsustainable. It's not agriculture that's inherently destructive or extractive. Our agriculture is a reflection of our culture.

Since European settlers arrived on this continent, our conventional agricultural systems have been based on extraction and exploitation. The extraction of land through genocide of the Indigenous peoples. The exploitation of people and the extraction of labor through slavery. The extraction and exploitation of natural resources: topsoil, aquifers, fossil fuels, waterways, and forests. Modern industrial agriculture is built upon the belief that we have the right to take whatever we want without giving back in return; that we can manipulate the natural world as we wish without consequence. It requires a stubborn refusal to take the hints Mother Nature has given us—hints of increasing urgency that now include record flooding, record storms, record wildfires, record temperatures—and instead, doubling down on attempts to control and dominate her.

I used to believe that individual choices made by individual people would drive markets to change in response. I used to believe that if enough consumers prioritized the humane treatment of animals and sustainable agricultural practices when shopping for food, we could inspire change industry-wide and transform the system through our choices. This, in effect, is "trickle-up" change: decisions made at the level of the individual that eventually, once a certain threshold is crossed, trickle up to create systemic changes across an industry. It can be effective. Thirty years ago, when I was growing up, organic produce was available only at natural grocers or farm stands; now it can be found at big box chains. But I no longer believe this is a viable tactic against modern industrial agriculture, at least not within the time frame we have, based on overwhelming scientific consensus, to address and reverse our ever-climbing carbon emissions. In 2018, the Intergovernmental Panel on Climate Change released its special report stating that, in order to prevent runaway climate change, we need to cut global net emissions in half by 2030. Individual choices have profound, even life-changing impacts on the people making those choices and the farmers and ranchers they are in relationship with. But I've come to the conclusion that believing trickle-up change will bring about the seismic shift we need in the time frame we have is as naive as believing trickle-down economics was ever meant to help anyone but the upper class. To restructure our entrenched agricultural systems, at scale and at speed, policy must change.

What if we restructured agricultural subsidies to support tactics that help solve the current problems we face? Since our tax dollars are at play, doesn't it make sense to use them to encourage practices that benefit us all? What if, instead of subsidizing specific crops, we subsidized sequestered carbon? Carbon, drawn from the air by plants through photosynthesis and sequestered safely in the soil with the help of mycorrhizae and microbes, pastured animals, organic farming practices, and the people who steward the animals and the land. Nearly 80 percent of the total greenhouse gas emissions in the

United States come from transportation, generating electricity, and industry. What if agricultural subsidies were paid toward sequestering those carbon emissions in the soil? Under this model, crops themselves would not be subsidized, and the value of any given crop would be determined by the open market. Farmers would have the freedom to choose what crops made the most sense for them to farm, with financial incentives granted through the amount of carbon they sequestered in their soils.

The reciprocal relationship between perennial pastures and free-ranging animals that results in sequestered carbon isn't exclusive to beef cattle—dairy cattle, bison, sheep, goats, pigs, and poultry can all help fortify the land while nourishing themselves in the process. Nor is carbon sequestration exclusive to perennial pastures. Through organic and regenerative farming practices, cropland can sequester carbon in the soil as well. Even corn and soybeans can be raised while sequestering carbon, just not via the current industrialized method. As biologist Dr. Robin Wall Kimmerer describes in her book *Braiding Sweetgrass*, her Indigenous ancestors planted corn, beans, and squash together—one seed of each per square foot of soil. The corn stalk acted as a trellis for the bean plant; the squash leaves hindered weeds and reduced evaporation of moisture from the soil; the roots of the bean plant, with the help of soil microbes, fixed nitrogen in the soil to nourish all three. These three plants thrived together. They thrived *because* they were planted together. And, in this configuration, more food was grown on the land available than if each were planted in segregated rows. Modern farm machinery is not capable of harvesting corn, beans, and squash planted in this manner. Harvesting poly-cultures requires the careful work of humans, the eyes and hands of individuals gathering the treasures of the earth—more individuals than the one it takes to drive a combine, the one destined to be replaced by self-driving combines if we continue on the path we're on.

Subsidizing sequestered carbon would open benefits to all farmers and ranchers, regardless of what they chose to grow or raise, while incentivizing sustainable and regenerative practices. The specifics of those practices would inevitably vary by location, by crop, by climate, by person. Each microclimate, each species of plant, each plot of soil offers different strengths and constraints, has different needs—working with nature is not one-size-fits-all. What remains true across all terrains and microclimates and crops is that carbon can be seques-tered safely in the soil through regenerative agriculture.

The scientists and researchers at Rodale Institute, who have been studying sustainable agriculture for seventy years, have concluded that "if we converted all global croplands and pastures to regenerative and organic agriculture, we could sequester more than 100 percent of current annual emissions." Aggregated data from studies around the world have shown that regenerative farming and grazing can seques-ter the equivalent of two to ten tons of atmospheric CO_2 per acre of land each year. And this can continue for decades—some studies esti-mate up to a century—before the soil reaches a "saturation point" with the amount of carbon it can store. At that point, the soil doesn't se-quester additional carbon, but it can store the carbon already seques-tered for thousands of years. In 2018, CO_2 emissions from the United

States reached 5.4 billion metric tons. With close to a billion acres of farm and rangeland in the United States, including about two hundred million acres of BLM pasture leases, we have, literally under our feet, an abundant, powerful ally in reducing our net emissions, one that can be utilized immediately: our soil.

If the heart is the center of a body, then perhaps the prairie is the heart of this country. For millennia, the prairie covered nearly a third of what is now the continental United States. The spectacular biodiversity of the American prairie is exceeded only by that of the Amazon rainforest. Short-grass prairie stretched from west Texas to central Montana. Mixed prairie reached from central Texas, through western Oklahoma, Kansas, Nebraska, and the Dakotas. The tallgrass prairie, profoundly fertile and abundant, boasting grasses that grew eight feet tall, spread from eastern Texas up through Oklahoma and Kansas, across much of Nebraska, to eastern North Dakota; and covered the entire state of Iowa and much of Illinois, Minnesota, and Missouri. Today, only 3 percent of our prairie survives. Of the tallgrass prairie, just 1 percent remains. The tallgrass prairie is one of the rarest and most endangered ecosystems on the planet, and the most endangered in North America.

What was once tallgrass prairie is now covered in monocrops of corn and soybeans. If sequestered carbon were subsidized instead of specific crops, perhaps some farmers—perhaps more than some—

would determine that the most effective way to sequester carbon in the soil is not through crops at all, but through allowing the land to return to native tallgrass prairie. With the return of the prairie, the abundance of life that creates the prairie, that is supported by the prairie, would also return: bugs and bees and deer and voles and lizards and hawks and owls and fox and wild grasses and wildflowers and wild herbs and wildlife. The perennial prairie is like a quilt pieced together by Mother Nature, incorporating dozens of species of grasses, hundreds of species of plants, and countless species of mammals, reptiles, amphibians, insects, birds, and fungi.

Model simulations out of UC Davis demonstrate that grasslands are a more reliable carbon sink than forests, for grasslands are less vulnerable to drought and wildfire. Even when fires sweep through grasslands, the living plant and roots of perennial grasses can survive belowground and emerge again. Grasslands and ruminants evolved together—they depended on one another, supported and enhanced one another, and flourished together. If monocrops of corn and soy returned to prairie, the prairie would need ruminants to thrive. We no longer have the great herds of buffalo that once roamed the prairie, though their populations are being fostered. But we do have cattle, sheep, and goats.

A common argument in defense of CAFOs is that we don't have enough land to produce our current supply of meat any other way. In a published study, Dr. Jude Capper determined that, to maintain our current volume of beef production, and to do so exclusively on pasture, the amount of additional grazing land required would be equivalent to 75 percent of the land area of Texas. It so happens that 75 percent of the land area of Texas is about 125 million acres, nearly 60 million acres *less* than the land currently planted to corn and soy. What if farmers returned that land to prairie, to the vibrant ecosystem it is meant to be? And grazed it with yearling cattle bought from ranchers—for yearlings on pasture practically raise themselves—or leased the land to would-be ranchers who don't have land of their own, or started their own herds? The land could return to the most fertile, productive grassland on the continent, capable of supporting tens of millions of large ruminants. And not just cattle—multiple species can be rotated on the land. Chickens and turkeys are excellent allies for insect management; sheep and goats prefer forages that cattle won't touch. With cooperation among species, people, and policy, it's likely there would be no decline in the amount of meat available on the market. And while this metric is important, it's just one marker of success. Topsoil would begin to be restored, not lost. Ag-based pollution and runoff, and the amount of fossil fuels required for food production, would drastically decrease. Instead of being vulnerable to flooding and drought, this carbon-rich soil, stabilized by perennial root systems, would hold water like a sponge. Beyond our domesticated animals—in part, because of them—the exquisite biodiversity of this region, long banished, could finally return.

Ironically, resiliency in nature and resiliency in our food systems are no different from resiliency in financial planning—"a diverse portfolio" ensures that one glitch, one failure, won't wipe out everything. The latter is common knowledge among people in skyscrapers and capitol buildings who oversee billions of dollars; it should not be hard for those involved in agricultural policy to comprehend that ecological systems, and thus, agricultural systems, are no different. More diversity leads to stronger, healthier, more resilient systems, from the soil itself to the food we eat, to our communities, our country, our culture.

When animals are allowed to roam and graze as their bodies are designed, illness is the exception, not the rule. Mike's oldest cows, free to roam, free from antibiotics, fed only grass and alfalfa, have lived to twenty-one years old. In CAFOs, illness is such a common and expensive threat, antibiotics are given prophylactically. Bovine respiratory disease is the leading cause of illness and death in CAFOs, with the majority of cases falling within the first six weeks that cattle spend in a CAFO. Bovine respiratory disease in CAFOs costs an estimated $800 to $900 million each year due to the cost of treatment, decreased weight gain in sick animals, and death loss. While the mortality rate in CAFOs is considered small, that rate has climbed in recent years. It held around 1 percent through the 1990s and crept up to 1.5 percent by the late 2000s, then doubled between 2010 and 2014 to reach an average of 3 percent death loss. If that trend has continued, death loss in CAFOs could currently be closer to 5 percent. With about twenty-nine million cattle confined in CAFOs annually, a 3 percent death loss amounts to around 870,000 dead calves each year, most dying within their first month in confinement. A 5 percent death loss is almost 1.5 million animals. That degree of waste—of life, first and foremost, but also of potential food, of resources, of hay and feed, of time and work spent tending the mother cows during gestation and calving, and caring for the calves before they enter CAFOs—is unconscionable, especially since it has been shown, over the course of decades, that a significantly lower rate is achievable. Stress, travel, and diet have been determined as primary factors leading to illness and death in CAFOs—when it's costing this much annually, serious research has been dedicated to the issue. And yet, instead of directly addressing stress, travel, and diet, CAFOs attempt to mitigate the problem with antibiotics and administer growth hormones to artificially balance out decreased gains due to illness and stress.

Without oversupply depressing the price of corn and soy and subsidizing CAFOs' bottom lines, could CAFOs stay competitive? With sequestered carbon subsidies fortifying the income of farmers and ranchers, and with animals grazing and foraging the restored prairie, the quality and nutritional profile of meat, dairy, and eggs would increase while the consumer price of these organic, pastured animal products would likely remain consistent with the current price of those from CAFOs. The same would be true of regeneratively farmed fruits and vegetables—the quality and nutritional value of produce would skyrocket, while prices would stay similar to those of current conventional crops. Imagine if the food most widely available to everyone—because it was the majority of the food on the market—was healthy, organically raised, and nutrient dense, and not only nourished our bodies but also helped protect the planet.

Responsibly grazed rangeland sequesters carbon. Mob grazing can do so even more profoundly. Also known as rotational grazing or managed grazing, mob grazing is a form of biomimicry that entails moving cattle often, bunched in a group, between numerous smaller paddocks built within a larger pasture to emulate the grazing patterns of wild herds. As prey animals, vulnerable to predators, wild ruminants traveled in tight groups and moved constantly. Mob grazing replicates the beneficial effects these tightly packed, transient herds had upon grasslands, increasing the vitality and production of the land by as much as 45 percent. Some ranchers who practice mob grazing move their cattle (or sheep, or goats) every three days. Others move them three times a day. This is distinctly different from the passive grazing of cattle scattered across one large pasture for an extended period of time. Mob grazing has been shown to dramatically improve the health and fertility of the soil, markedly increasing its ability to sequester carbon and store water. Mob grazing can reverse desertification and restore barren land to a flourishing ecosystem.

Contrary to popular belief, undergrazing land can be just as harmful to plant and soil health, and thus, the entire local ecosystem, as overgrazing. The key is not in leaving the land alone, but in mimicking the systems that evolved in that place. Mob grazing is a remarkable example of humans and nature working in tandem to mutual benefit. Rancher Joel Salatin, who has practiced rotational grazing for decades and whose pastures have become so lush they support five times as many cattle as his neighbors', has declared, "If every farm did this, we could sequester all the carbon released in the industrial age in less than ten years."

Mike and I don't practice rotational grazing. The desire is strong, and moving cows is the easy part—cows learn routines so quickly. But setting up the system takes a significant investment in time and money: cross-fencing must be built, miles of electric fencing and solar-powered chargers bought, water systems developed so the cattle have access to fresh water from every paddock. It's been hard to take the leap when we don't know if we'll have our current pasture leases the following year. Even for those who own ranchland, the logistics can be daunting. Mob grazing can take more time and require more infrastructure; though, perhaps most daunting is the learning curve. Pasture size, and the distance between pastures and a rancher's home, varies by region and by ranch. A two-hundred-acre pasture in New England might run the same number of cows as a two-thousand-acre pasture in New Mexico. While rotational graz-ing has been shown to be profoundly beneficial in a wide variety of regions and climates across the world, implementation can be more challenging for some.

If agriculture is to become sustainable and regenerative on a large scale, there must be a financial incentive to help this shift take place. Some people are motivated only by money. Corporate farms that prioritize shareholder returns may base all their decisions on the bottom line, regardless of the impact on lives downstream. But for others, I would venture most, profit margins are simply too small to comfortably allow room for experimentation. Change is terrifying when margins are tiny, when one mistake means you could lose everything. Ranchers transitioning to mob grazing, and farmers transitioning to organic practices, to entirely different crops, to polycultures where they once raised only monocrops, or to incorporating animals to reduce the need for synthetic fertilizers, face considerable challenges while actualizing these changes. Financial incentives and sequestered carbon subsidies would help make these transitions more feasible.

Subsidies and incentives created massive corn and soy monocrops and laid the foundation for CAFOs and destructive industrial agriculture. We created this system. And we can change it. Restructuring subsidies to fund sequestered soil carbon instead of specific crops would support those doing work that benefits us all. These financial incentives would pay dividends in the form of more widely available nutrient-dense food, a healthier and more vibrant environment, and offsetting our collective carbon emissions through regenerative agriculture.

They say, at the end of it all, there will still be cockroaches and coyotes. I don't know much about cockroaches, but I've been living with a coyote for over a decade and I've learned some of his secrets. Coyotes are survivors because they are hyper-aware of changes in their environment, and they adapt swiftly and deftly. They're not attached to the past. If the present demands change, they change on a dime. When discussing his book *The Uninhabitable Earth*, David Wallace-Wells stated, "The main driver of climate change is *what we do*—how much carbon we put into the atmosphere—and ultimately, that's a reminder of how much power we have over the system." Sometimes, I wonder how differently I might make decisions if I woke up one morning with amnesia, my mind wiped clear of my past experiences and my stockpile of reactions and attachments to those experiences. It's clear what I would lose if I suddenly started from scratch, but I wonder, sometimes, what I might gain.

We have two options: we can continue extraction expansion—mining, fracking, and drilling—and maintain our dependence on fossil fuels to raise monocrops, with the bulk of our farmland going to corn and soy for export, ethanol, CAFOs, and processed food; or we can decide to use our creativity, our will, and yes, our money, to move agriculture in a more productive and sustainable direction. Not just sustainable, but regenerative—to return agriculture to an act of nurturing, a multidimensional labor of love that nurtures the land and all the elements, the planet and all her inhabitants, our bodies and our health.

I've been present for more births than I can count. As anyone who has given birth or witnessed birth knows, birth is not delicate harp music and feather beds. Birth is shit and blood and slobber and sweat. Birth is *messy*. It's so beautiful and so messy. As it begins, the act of birth seems so implausible, if not wholly impossible. It seems there are so many ways it could fail. If you didn't know what was coming, you might be convinced that the mess, the discomfort, the challenge, the risk, couldn't possibly be worthwhile. And then, with singularly profound engagement and focused effort, new life is born. It exists! And sure, that new life falters before it gets its footing, but once it does, it's off and running. To bring something new into the world, we have to be willing to get messy. It can be overwhelming to contemplate the restructuring of our systems, the work and the mess involved. To have faith that the apparent implausibility is, in fact, an illusion. Birth is a revolution unto itself, an invitation to surrender to the discomfort, to embrace the mess, to transform ourselves and everything we know for the promise of the new.

III

———⟨∞⟩———

Love does not consist in
gazing at each other,
but in looking outward together
in the same direction.

—ANTOINE DE SAINT-EXUPÉRY

CHAPTER 8

An Unsteady Faith

would have spent every summer on the mountain if I could, but once I started selling beef, the logistics of the business demanded I stay in the valley. When people continued to show up at my door over the following summers, I knew I had to move. Mike and I were ready to take the next step of living together, and his house, situated well off the road, offered more privacy. The cabin had been my home for six years, and despite its challenges, leaving it was devastating. I couldn't bear to completely move out. I left every nonessential item I owned in the cabin—clothes, books, art on the walls. If I needed a particular book, I went to the cabin to get it. When the snow arrived and I wanted my corduroy pants, I went to the cabin to get them. Little by little, piece by piece, my belongings migrated to our new home.

The move was not purely heartbreak of course. Away from the

activity of the road, Charlie became more calm and relaxed, more content than he'd been since our summer on the mountain. Mike and I transitioned seamlessly to living together. And I marveled at the luxury and convenience of a "normal" house. A sink, inside, at which to brush my teeth! Heat from a button on the wall! My wonderment continues to this day. In the scope of human experience, across the world and across time, these are phenomenal luxuries.

Just after he turned four, Frisco got sick. That winter was unusually cold, the snow so dry it kicked up into glitter with every step. Frisco lost his appetite, and about 250 pounds with it. His weight loss was severe and frightening, yet his eyes remained bright and his spirit lively. I made him a nest in Daisy's milking barn and covered him with an unzipped sleeping bag, which he wore like a horse blanket. I got sick, too, but that didn't keep me from throwing on layers of silk and wool the moment I woke up and heading down to the barn, still groggy from sleep, snapping awake the moment my face hit the frigid air. I stayed with Frisco for as much of each day and afternoon and night as I possibly could. My vet came out, took blood samples and stool samples, warned me to prepare myself for the worst. He left me with every treatment he thought might help—pills the size of a roll of quarters, syringes that were even bigger.

I dug through the haystack to find the best grass bales, hoping the tender, fragrant hay would entice Frisco to eat more. I pitchforked the wet, shitty straw out of the barn and refilled Frisco's nest with clean, gleaming straw the color of sunshine. I talked to him, sitting in the straw beside him. I stroked his soft, giant neck, his cheeks, his forehead. I curled up beside him as he lay in the straw, and listened, with tears in my eyes, to the intermittent squeaks and creaks of Frisco grinding his teeth, one of the few ways cattle express pain. I got filthy every time I went down to the barn, from hay and straw dust and mud and shit and sweat, but the grimiest feeling came from the layers of dried tears on my face; salty, scratchy on my skin. When I returned home, I peeled off my clothes and headed straight into the bath. In that warm cocoon of water and steam, I cupped handfuls of water over my face to dissolve the tears and allowed myself to rest, to let go.

April arrived and winter still clung to the landscape, but the sandhill cranes returned, promising, with their otherworldly song, to call in spring. The midday sun was not warm, but it was bright, and Frisco wobbled out of the barn on unsteady legs. He stood broadside to the sun, soaking up the rays with Daisy and Fiona. He ate some hay. He drank some water. I smiled. It seemed to be a turning point. Frisco lay down in the sun, and I sat on the ground beside him, leaning against his massive shoulder. I stroked his cheek, timed his pulse with my phone. Frisco closed his eyes. And then he swooped his head around so that I was tucked in the crook of his neck, and he laid his head upon the full length of my torso and fell asleep.

The next morning, Mike got up early to check on his cows. They had just started calving, and the freezing nights were dangerous to newborns. I woke up with him. Mike offered to check on Frisco on his way out, and because it was so dark and cold, and because I was so tired, I agreed, and asked him to call to let me know how Frisco was. Five minutes later, Mike called to say that Frisco was awake, that he was lying down but his head was up and his ears were perky. I felt relieved and overjoyed as I did every morning when I confirmed that Frisco was alright, that he was still with us, that I was going to have another day with him. I pulled myself out of bed. An ever-growing pile of clean but unfolded laundry sloped across my dresser and onto the floor. *I might as well fold this laundry while I get dressed*, I thought to myself. I folded it and put it away, and from that domestic task, I turned to general tidying which had been neglected for the past month. I heard Mike's truck pull in. I looked at the clock, for it seemed he had not been gone long at all, but I had lost track of time, and it had been just over an hour. Mike appeared in the doorway, stricken. "Frisco didn't make it," he said. "What?" I said. "Frisco didn't make it." I did not understand. Mike kept repeating those words, "Frisco didn't make it," and I stood in the center of the room repeating, "What?"

"I stopped by the barn on my way home," Mike said. "I'm so sorry. He didn't make it." Mike crossed over to me and put his arms around me. I stood with my arms at my sides, my ears ringing and my head spinning, like when I got out of the bath too quickly. I squirmed away from Mike. "I have to go down there," I said, and ran to the barn.

The moment of his death couldn't have been more than fifteen or twenty minutes before I got there. He was still so warm. His eyes hadn't yet disappeared. I laid against him, felt the heat from his body on my face though my insides felt like dry ice. He had died alone. He'd been alone while I was folding laundry. I was so tortured by this I could barely breathe. And I tortured myself further by wondering what I would have done if I had been there. Would I have tried to resuscitate him? Would I have been completely hysterical? Would being there have been worse than not being there? For him? Or for me?

We buried Frisco on the point of the hill overlooking the pond and the grassy draw. Mike dug an enormous hole in the frozen ground with his backhoe. Then he loaded Frisco's body on the back of the feed truck using chains and the hydraulic steel arms on the back of the truck used for loading and unrolling sixteen-hundred-pound hay bales. He drove to the grave where I sat and lowered Frisco's body into the ground. As Mike filled the grave with dirt, the metallic clang of the backhoe shovel echoed through my bones and shattered my heart. I couldn't look at Mike while he worked, or even after he was done. Acknowledging Mike's grief only amplified my own, and my own was too much to bear. As Mike drove the backhoe back down to the corral, I watched a calf being born in the morning sun in the draw at the edge of the pond below. It was not the instantaneous reincarnation of Frisco. Frisco was gone. I lay in the soft dirt of Frisco's grave and folded myself up like a wing.

It was close to 9:00 a.m. when I finally went inside and made tea and topped it off with rum. I don't drink often enough to keep a supply of alcohol in the house, and Mike doesn't drink at all, but an unopened bottle of rum had sat in a cabinet for two years, left by one of Mike's hunting pals. It remained unopened after all that time because I don't like rum. But that morning, escape trumped flavor, and I drank mug after mug of tea spiked with that old bottle of rum. Frisco had been my compass. He had defined my days and my thoughts for months. He was my first calf, my giant, my friend, and now, suddenly, he was gone. I once read a line somewhere, I can't remember where, but I wrote it down at the time: "Being prepared for the unexpected means being prepared to be unprepared." I wasn't prepared. I wasn't prepared for Frisco's death, even after months of illness. Even as he lost weight, even as he became increasingly unsteady on his feet, I believed Frisco would get better. But I had been incredibly superstitious. I refused to publish any pictures of Frisco in which he was backlit, any images of him that had flare or in any way looked ghostly. Perhaps I was superstitious because I already knew, because I saw in my photos a foreshadowing of the future. Perhaps it wasn't superstition at all, but a silent scream against what was happening.

By midday, the escape I longed for still eluded me. I wanted to be with Frisco, and the only way I could do that was to go back out to the point of the hill. The rust-red dirt of the new grave was as dry and fine as flour, and I lay spread eagle across it, Frisco's body in the earth beneath me. I wriggled into the dirt until I was half buried myself, drunk and sobbing, and stayed there until the cold crept through my down jacket and the numbing sorrow and the alcohol daze.

The next day, I couldn't see colors. I looked out the window and saw a clear sky above the great expanse of open land that stretched to the mountain, and my brain knew the view burst with radiant orange earth against a deep blue sky against snow white peaks. My brain knew I was gazing at the exhilarating vibrancy that had always given me such pleasure to behold. I knew I should see it, but I couldn't. It looked gray and dulled, like when the smoke blows in from California

wildfires and everything—colors, distance, the edges of things—
becomes muted from the haze.

In mourning, in grief, I traveled to an in-betweenland. I found
myself in a realm between worlds, where the veils seemed to drape
right across my shoulders and slide like silk across my face. The familiar world seemed hazy and blurred, as if from across an expanse of
altitude or drugs. This world, our world, felt untouchable, but the
realm of the departed was untouchable, too. Frisco was untouchable,
unreachable, and a part of my heart became unreachable, too.

Trying to understand death is like chasing rainbows. There's no
ambiguity in death itself. But even in its obviousness, its undeni-
ability, death seems impossible to grasp. We have so many theories: of
heaven and hell, of nirvana, of reincarnation; of joyful reunions, that
these endings are not permanent; of returning to source, of returning
to soil, of becoming stars, of nothing more or less than becoming fod-
der for a blade of grass. I'd had my own ideas, born of acid trips and
mysticism, but when Frisco died, all my hopes and theories shattered
at my feet like broken plates. Belief and faith were no longer enough—I
wanted proof. Without proof, our collective hypotheses of life beyond
death seemed like nothing more than fairy tales, stories we tell ourselves to keep from going insane.

Before Frisco died, I wondered if death was perhaps another birth
process, the act of being born into another plane, into a reality we can't
fathom. If we knew our loved ones were being birthed, via death, into a
realm of endless joy and unadulterated connection, it would be easier to
let them go. If I knew that death was as glorious and profound as birth,

worthy of annual celebrations of that day ever after, I would never want to drag my lost loves away, would never daydream of pulling an Orpheus, succeeding where he failed, and bringing them back to me. But we don't know, and that question mark became a dagger in my heart.

Hardest for me was the possibility that Frisco ceased to exist. That death is the end, period; not a transition to another way or place or manner of being. That he was over. That he would experience nothing further. That he was done. It was sad enough that he was no longer here to watch the moon rise or walk in the snow or eat something delicious or laugh with or hug. I've never feared death for myself (even if it is a complete ending, total doneness), but the older I get, the less ready I am to say goodbye to the simple wonders of life. I couldn't bear the thought of Frisco's existence being over. I wanted evidence to the contrary. I wanted to stand in the multicolor light of truth and claim the ultimate treasure: losing the loss by finding a bridge between worlds or realms or realities or time so the separation between us no longer existed. But, like rainbows on the horizon, the more I searched, the more elusive that became. I found no rainbow's end, no portal, no proof. All I found were more questions. Death is the great unanswerable. We can't truly know the other side of death, if there is another side of death, what happens next or if there is no next, until we die ourselves. And then, I suppose, the answer doesn't matter anymore.

And yet . . . and yet. I've seen so many dead bodies in my work on this ranch and as an EMT, and I've seen what's missing, I've seen what leaves. Actual dead bodies don't look like the dead bodies we see in movies. Dead bodies in movies are, of course, living people, and a truly dead body looks more like an abandoned pile of clothes on the floor, even if that body is naked, even if that body is an animal and clothes were never part of the equation. A dead body is almost unrecognizable from the living being it had been. It looks like a container that has been left, a shell discarded, like incredibly ornate Tupperware that had held something else, something more, be it spirit or soul or light or god. It is only from seeing what leaves, seeing the vacancy with my own eyes, that buttresses an unsteady faith in what we cannot prove.

The stage of grief they don't tell you about—and this may only be if the one who died was a dependent—is the paralyzing self-doubt. The overwhelming conviction that everything you do henceforth will be wrong. That every choice you make will be a mistake. That you will not be able to survive the consequences of your actions because, when you thought you were doing everything right, when you thought you were doing everything you could, they died! They died. After Frisco died, I found myself bent up in agony from paranoia—not a paranoia of the outside world, but an internal paranoia, a paranoia of self. Why was every phone call so hard? Why was I choosing inaction over any action whenever and wherever possible? Why did I have such an epic distrust of myself? What could I trust if I couldn't trust my intuition? Why was I so scared all the time? How the hell could I run a business like this? The subtext of each question was the one question that tormented me: what could I have done differently to save him?

Coping with death is like a wound; it heals like a wound, slowly, over time. And if it heals in a healthy way, like a wound, it heals from the inside out. If all that heals is the surface layer, everything beneath will rot. It dawned on me how happy I had been through the winter and early spring, despite the agonizing uncertainty of Frisco's lingering illness, despite the cold, despite the stress. I was happy because I was with him. It doesn't matter how weak our loved ones become, it doesn't matter how much help they need. It doesn't matter, as long as they're here. Now, joy was a speck in my rearview mirror. I felt so far away from "happy" that I didn't know how to begin to find my way back to it.

Charlie didn't cure me of my sadness, but he made me smile on days I could smile, and he lay beside my bed on days I didn't bother getting up, and he spun his coyote magic around me like a cocoon. I marveled at Charlie every day, and this, perhaps, is what saved me. Marvel is an interesting phenomenon. It is possible to marvel even in the depths of sadness. Marvel doesn't require joy or happiness—it just requires noticing. I marveled at the colors of the clouds at dusk, at the calls of the sandhill cranes echoing off the hillside above the pond. I marveled at Charlie's beauty, at the gentleness of his kisses and licks, at his presence beside me as if he knew how much I needed that. I marveled at my ability to marvel and it made me realize I was not completely lost.

And though I was so sad, I was grateful every day, which also does not require happiness. I was grateful to have a job that allowed me to hide in bed twenty hours a day without getting fired or losing all my income. I was grateful for a house with a furnace rather than just a woodstove so I didn't have to get up and chop wood; grateful for the bathtub and indoor running water; grateful for Mike's patience and understanding. I was grateful for Charlie and Chloe, and grateful that they were willing—even happy—to pile on top of me and sleep the days away.

On one of those days, when I was napping around noon, I woke to Sir Baby bellowing like a foghorn. The sound rumbled through the walls of the house like the Jake brakes of a semi. When Baby's roars lasted for more than twenty seconds, it usually meant there was another bull around, and each was telling the other how superior he was. A neighbor's bull had shown up at our back fence a few times the previous week, and we did not want him in with our cows. I rolled out of bed, slipped on my Vans, and went outside to see if he had come around again. The wayward bull was nowhere in sight, but once outside, I heard a cow mooing like I'd never heard mooing before. It was a cross between a desperate scream and an eerie, furious wailing. Sir Baby's booming alarm had obviously been in response to this noise. I walked around the corner of the house to see what was going on. A heifer had just calved—afterbirth still hung from her body. She loomed over her newborn calf and was pummeling it, head-butting it into the dirt while mooing wretchedly. I broke into a run and raced up the hill to get between the heifer and her calf.

I was completely nude but for my Vans. I had planned to peek out my door for the roving bull and get dressed only after I'd determined he was present; otherwise, I was going back to bed. But with a calf in mortal danger, there was no way I could waste time putting on clothes. The calf was limp and floppy, born just moments before I got there. I stood over the calf and flapped my arms in the heifer's face to shoo her away. She feinted to one side, dodged my arms, and bit one of the baby's legs. As she turned her rage on me, I grabbed the calf around its rib cage, clutched it against my body, and ran. The calf was heavy, and slick against my bare skin; so slippery with amniotic fluid, I couldn't keep a firm grip around its body. The heifer charged after us, ramming the calf with her head when it slipped from my arms. I kept running as fast as I could, half dragging the seventy-pound calf across the pasture and the driveway. When I reached the fenced front yard, I dropped the calf on the ground beneath the bottom rail, leaped over the fence, and pulled the calf under the rail and into the yard just as the heifer stormed up. She paced the fence and mooed furiously. I collapsed on the grass beside the calf, gasping for breath, covered in birth slime and dirt.

When I caught my breath, I went inside, threw on one of Mike's shirts, and grabbed a towel for the calf. I sat beside the calf in the sun, dried him off, and checked his body to see if the first traumatic minutes of his life had left him with any injuries. Calves are incredibly resilient and he seemed unharmed. Soon, he wobbled up to standing and nuzzled against me, looking for an udder. We needed his mother for that.

The calf and I hung out in the yard, and when Mike got home, I filled him in on the mama drama. Most cows will follow their calf anywhere in the hours and days after they calve, making it easy to move the cow, if need be, by leading them with their calf—holding the calf while riding a four-wheeler or pulling the calf in a child's sled. This new mother wanted nothing to do with her calf, so I got Daisy to help me lead her to the corral. With a curry comb in hand and a few slaps to my thigh, I called Daisy over and she set off behind me down the driveway toward the corral. Mike circled behind the heifer to encourage her to follow Daisy, and once he saw the three of us rounding the hill to the corral, he rode down with the calf on his four-wheeler.

At the corral, Mike and I used the head catch to keep the heifer stationary so her calf could have a chance to safely nurse. A head catch is a contraption of two metal panels that open and close with a big lever, and these panels gently close around a cow's neck without actually putting pressure on the cow. A cow's neck is much narrower than her head and her shoulders, so when the panels are in place, she can't move forward or backward. Since the panels aren't squeezing her neck, she can move her head up and down, and even eat while standing there. Head catches aren't traumatizing to a cow as long as the person running the levers doesn't let the panels slam against the cow or otherwise abuse her. With the heifer in place, placated with hay, we maneuvered the calf to her side. The calf latched onto a teat and nursed with gusto. The heifer didn't kick or get impatient while her baby suckled, and when the calf finished his meal of colostrum and toddled to the front of the head catch, the heifer, who was still confined, began licking him. Just as Mike and I smiled at each other hopefully, the heifer changed her mind and head-butted the calf right in the ribs.

It was too dangerous to leave the calf with his mother overnight. We kept them both in the corral, in separate sections, with just a rail fence between them so they could smell each other and communicate. When we put them together the next morning to see if the heifer had decided to love her calf, she went right into attack mode. Back in the head catch she went, so her calf could have his breakfast. Afterward, we separated them. We brought them together midday and later that evening, and again the following day. Each time, there was no change in the heifer's baffling behavior. Having to depend on the head catch two or three times a day for the calf to nurse was going to be high maintenance. Moreover, we didn't want the calf to associate food with us and the head catch, and have that association override his instinctual connection with his mother. So, on the fourth day, we hobbled her.

Hobbles are like leg shackles made of leather that go around a cow's hind ankles, though they're more commonly used with horses. Hobbles allow an animal to walk with very small steps, but hinder the speed of movement. We hoped the hobbles would keep the heifer still enough for the calf to nurse and make it harder for her to quickly turn and head-butt him. It took mere seconds for the heifer to figure out how to bunny-hop with the hobbles on, taking a giant step with her front legs and jumping with her back legs together. When the calf approached her, the heifer pivoted, whacked the calf, then bunny-hopped at extraordinary speed to the far end of the corral.

We took off the hobbles and devised another contraption to protect the calf. We clipped two horse cinches together with a carabiner and buckled them behind the heifer's shoulders. I put a halter on the heifer and tied a lead rope between the chin strap of the halter and the bottom of the cinch between her front legs. She could still eat and drink and look side to side, but didn't have the full range of motion to head-butt with any force. Since none of our cattle have ever worn tack, Mike was nervous about an epic protest of bucking and snorting, but the heifer stood calmly beside me like a seasoned saddle horse as I fiddled with the halter and quick-release knots.

With this contraption in place, we brought the calf to her side. Though the heifer could no longer head-butt her calf, she wouldn't stand still to allow the calf to nurse. She walked a few steps away whenever the calf approached. Mike shook his head, unbuttoned his shirt, and tossed it over the heifer's face. Unable to see, she stopped midstep and stood completely still. She didn't panic or try to shake it off, and the calf snuck in and had his meal. I placed an armful of hay in front of the heifer's nose to relax and distract her.

We couldn't leave the heifer blindfolded, but we still didn't trust her with her calf. Between meals, we separated them in adjoining sections of the corral, and brought them together two or three times a day. We often found them lying side by side, with just the rail fence between them. At meal time, either Mike or I, whoever was on duty, tossed a flannel shirt over the heifer's face and tied the arms under her chin, then opened the gates for her calf. The calf galloped to his mother and latched onto a teat. The heifer seemed unfazed by the blindfold. Often, she stood and chewed her cud beneath the shirt, proof of utter relaxation. As the days went by, I began taking the flannel off midway through the nursing sessions to observe the heifer's behavior. Sometimes, she got antsy and angry and I put the flannel back on; other times, she stood calmly and nuzzled her calf. We began leaving them together after meals and watching their behavior. Once we felt we could trust the heifer not to hurt her calf, we left them together overnight. The next morning, I walked down to the corral at first light and caught the heifer standing peacefully, allowing her calf to nurse. It just took a little time, a little darkness, and plenty of trust.

Fiona had grown as tall as Daisy, with a muscular build that was all Sir Baby. But her demeanor was wholly her own. Fiona was as gentle and sweet as the scent of lilacs in the sun. She allowed me to lie on her back, just as Sir Baby did. Her coat was the exact color of the red dirt, with blond highlights that edged her ears and rimmed her eyes. Her udder was covered in soft, pale hair almost as light as Daisy's. Her udder had been steadily growing for the past few months, for she was pregnant. By the time the back wrinkle filled out, Fiona's udder was slightly larger than the average Angus cow's, though nowhere near the size of Daisy's. I didn't expect to milk Fiona much, if at all, but that would be determined by Fiona's body and her calf's appetite.

One mild morning, I noticed Fiona leaving the herd with her tail cocked, early signs of labor. She walked down the hill to a secluded clearing at the bottom of a large draw below Mike's house. I followed her down and brushed her neck, her belly, her back, her flanks; taking my time, feeling her relax under the brush. Then I went back up to the house. I tried to work, but broke away often to check on Fiona from the deck with binoculars. When I focused on the glistening tips of hooves just emerging, I jogged down to the draw. I wanted to watch the birth, but didn't want to bother Fiona, so I nestled myself in the grass at the opposite end of the clearing. Fiona crossed over to me. She stopped in front of me, dropped heavily to her knees and then her flank, and lay down beside me. She inhaled deeply and stretched out her body during a contraction. Her back hoof rested against my ankle. She lay still for a moment, sighed deeply, pushed again, and delivered her calf right into my lap.

Fiona lurched to standing, murmuring soft "mama moos" to her calf as she began licking her off. I stared at the calf's long, perfect eyelashes, ran my finger along a shiny hoof, remembering when Frisco was that tiny. My chest was tight; I felt too small, too inadequate to hold such awe and such loss simultaneously. As I wiped away tears, I caught sight of Sir Baby approaching from the corner of my eye. I didn't want him to interfere before the calf had stood and nursed. I got up to intercept him. On my way over to Baby, I saw a long, straight branch that had blown off a tree during a spring windstorm lying beside the trail, and stooped to pick it up. When I got closer to him, I held out the branch, horizontally between us, eye level to Sir Baby. He stopped his advance. The branch looked like a fence rail. I took a few steps toward Sir Baby, still holding the branch between us, and he started backing up. Sir Baby had the strength and power to bash through me and my flimsy stick, or he could have simply walked around the end of it. But he believed it was a fence and that he was powerless against it. It made me wonder how many of the barriers we believe are insurmountable in our own lives are nothing more than sticks held at eye level.

Sir Baby got bored staring at my faux fence, turned with a huff, and walked back down the draw and around the hill, out of sight. I turned back to Fiona. Her calf was at her side, nursing. Fiona stood patiently for her baby, mooing a lullaby. I left them to bond in peace and walked back toward the house, then changed my mind, and sat down to watch them from the top of the hill. I reached my hand to the ground behind me to lean back and accidentally trapped a bee. She stung me. The prick was sharp and hot as the venom filled my hand. At first, the sting didn't cause pain so much as burning. The heat took over my entire consciousness, filling my hand and my heart. Instead of trying to hide from the pain or ignore it, I decided to face it, to enter into it. *Bee stings can be therapeutic*, I told myself. *Replace your fear with grace*, I told myself. And so I leaped into the pain like I would leap into a lake. I closed my eyes and exhaled and drifted down, down into the pain. Sinking, but it felt like floating.

The heat in my hand and my heart stretched and mellowed; transformed into a warm, heavy pressure; uncomfortable, but oddly comforting. It began to throb, the pain vague but severe, challenging to localize. I couldn't pinpoint the precise location or quality of the pain. It was an unfamiliar tenderness. I could only say it hurt, it hurt a lot, it hurt everywhere. When I noticed myself predicting impending doom—that it was going to hurt so much more later, that the itchiness would drive me mad, that the swelling would impede the use of my hand—I stepped between myself and my thoughts and said: "You don't know any of that yet. You don't know. You can't predict. All you know is right now and right now is survivable. Right now is less awful than the shocking first moment, that initial dagger. Stay right here. Stay right now. Watch the white butterfly fluttering among the wildflowers. You were watching her and you forgot the pain, didn't you? You continued to feel it, but you forgot it." My words to myself had diverged from the bee sting and wound back to coping with death. They continued: "Hold the pain close, wrap your heart around it. Hold it close. It's yours. It's proof of love. The pain reminds you that you are here and you have a choice to make. The pain can't break you. The pain shows you how much you have to give, and how much you have received."

The pain of the bee sting was heavy and warm and oppressive, like lying beneath layers upon layers of quilts. Eventually, in its own time, the pain settled down. Which is how grief feels after a while—it settles down. It's not gone. But it settles down. When it does, it's a relief, yet almost worse. Because during that fierce, initial grief, I still felt such direct intimacy with Frisco. The first stage of totally encompassing grief, the nothing-else-exists-except-you-not-existing grief, is an exquisite bond; an intense, beautiful bond of togetherness. The last long hug.

I knew I had to let go. I knew I had to stop searching for proof, stop chasing reunions in my dreams. I slid all the way down that muddy mountain of wondering, where the footing is unstable and the summit unreachable. All we know for sure is Now. All we remember is Here. After the pain, and in the pain, death reminds us to do our best, here and now, and love as much as we can. All we can do is love while we can. As much as we can. Over and over, no matter how much it hurts. I can hope there's more, and must live as if there isn't.

CHAPTER 9

Ritual Work

spent so much time with my eyes and my hands on Daisy, I could tell with a glance when something about her was off. I was so attuned to her "normal" that any deviation caught my attention as reliably as my own body's signals. In early January, when Daisy was twelve years old, I saw her trudging through the snow, alone, pausing at intervals to lick her rib cage. The sun was bright, the cobalt sky as deep blue as Daisy's shadow on the glittering snow. She held her tail straight, cocked away from her body. She was in labor. Her most recent calf, Roxy, had been born nearly two years earlier. It had taken a long time for Daisy to get pregnant again. I was expecting Daisy to calve around the same time that Roxy would be having her first calf—in the spring.

I flipped back through my calendar and found a note scribbled nine months earlier: *Sir Baby crashed through fence to get with Daisy.*

Daisy had acted like she was cycling in the months following their encounter, so I hadn't thought Baby actually bred her. But the math added up. I lured Daisy up the hill and into the snowy front yard so I could watch her closely. The temperature in the afternoon sunshine was 12°F, and it would only drop when the sun went down. Even if Daisy's labor went quickly and the birth went smoothly, overnight temperatures would be dangerous for a newborn calf, and I planned to bring the calf in the house for the night. Daisy would not be pleased, but a night apart was preferable to the risk that came with the cold. With Daisy in the yard, I could keep an eye on her from the house and avoid having to carry a heavy newborn across the ice-slick driveway. I moved a small trough to the yard and filled it with warm water, gallon by gallon, from the sink. Then I lugged a straw bale into the yard, cut the strings, and it exploded into a soft nest.

By nightfall, Daisy's labor had not progressed. The temperature dropped to 3°F. The moon rose, enormous, golden. It had been full the night before. I laid out quilts on the sofa, which sat beneath a window that looked out into the yard. If I cupped my hands to the glass, I could see Daisy. Moonlight bounced off the snow and off Daisy's white coat. It seeped around the edges of curtains and slid across walls. Full moons make it hard to sleep, but they make flashlights redundant, and on this night, that was a gift. I made tea and sat on the sofa, watching Daisy. An hour later, Daisy was lying down and chewing her cud. An hour after that, Daisy was up again and I was sleepy, so I went outside. Even with the full moon, layers of stars beyond stars sparkled above us. The coldest nights are always the clearest. It had dropped to 0°F, cold enough to make every yawn painful, the quick gulp of freezing air like a shot of fire in my chest. Daisy crossed the yard and nudged my arm with her forehead as if to say, "Won't you brush my moonbody in the moonlight?" I knocked snow off a curry comb and brushed her until a bout of uncontrollable yawning drove me back inside. I cuddled into the quilts on the sofa and set my alarm for an hour out. Deep sleep never came, and I woke to the low, resonant tones of an owl just before the alarm went off. I went outside to check on Daisy. She was lying in the straw, awake, calm, and still. She showed no signs of active labor. The owl spoke again. I followed the sound with my eyes and found its horned silhouette at the top of a tree behind Daisy. I returned to the sofa and set my alarm to go off every hour for the rest of the night.

Daisy didn't calve during the night, and by midmorning, she acted like she wasn't in labor at all. But she didn't show any desire to rejoin the other cows—she didn't moo to them, didn't stand impatiently at the gate, waiting to leave the yard. There was no question in my mind that she had been in labor the day before. Nothing was adding up. Nothing was going as it should. At noon, I called my vet to schedule an ultrasound. He was in the field all day but didn't want to delay seeing Daisy, so Mike and I drove her down to his clinic when he got in at 6:00 p.m. The vet determined that Daisy had miscarried. She was fully dilated, and the vet was able to deliver the fetus. It was premature, developmentally on track for an April due date if not for the miscarriage. She hadn't been bred by Sir Baby, after all. The vet gave Daisy medication to address her mild fever and to help stimulate delivery of the placenta. Though the fetus was premature, Daisy's milk had come in, and the vet urged me to start milking that night.

Daisy's miscarriage was shocking and sad, but I was grateful the vet was willing to see us so late and relieved that Daisy was alright. By the time we got home, the darkness of midwinter night had descended. We let Daisy out of the horse trailer and back into the yard. I gave her a pile of hay and sat at her side to milk her out. Daisy's milk would be undrinkable until the medication cleared her system, so I didn't bother with pails and milked onto the frozen ground. As I settled into the familiar position, my nose against Daisy, inhaling her scent, my sleep deprivation vanished. The intangible melancholy of miscarriage vanished. The invisible needles of ice in the air vanished. All that existed was the black night before moonrise and Daisy's breath rising and falling beneath my cheek. The muscles in my hands twitched with recognition after a year away from milking. A sheath of endurance enveloped me as I milked as quickly as I could. My cheek rested against Daisy's warm belly and the circle of heat where our bodies met was my universe.

I kept Daisy in the yard so I could watch her obsessively. I wanted to make sure there were no additional complications, and having her so close gave me a better chance of knowing if and when she delivered her placenta. I happened to be outside with her when she lay down and delivered the placenta, mooing the soft "mama moo" as she did, breaking my heart. I milked Daisy twice a day and dumped the milk for a week. In January in Wyoming, daylight didn't last for twelve hours, so I milked her at dawn—just before sunrise—and again at sunset. I hadn't milked in deep winter since Frisco was born. When milking in winter, there is the ritual of layering up: wool leggings, wool socks, wool thermal, and a thin wool sweater—anything bulky impeded freedom of movement. Over that, I wore a full-length snow skirt (as functional as snow pants but not nearly as noisy), a down hooded coat, a wool hat, a silk scarf, and Muck boots. Every item was necessary to ensure warmth while sitting in the snow in below-freezing air for nearly an hour. It felt like dressing for battle; my enemy, the cold. I could not let it win, could not let it beat me.

When I emerged from the house in the blue light of dawn, Daisy was standing at the deck waiting for me, rhythmic cloud-puffs of breath steaming from her nostrils. I quickly learned to find the balance between shuttling hay from the barn quickly enough that Daisy didn't get antsy with anticipation for her morning meal and leisurely enough that I didn't get sweaty beneath my layers before milking. Sweat, cooling while I sat, chilled me far more quickly than the icy air. With hay for Daisy placed so that Daisy would stand in a particular spot that allowed me to sit on a granite boulder in the yard, I milked. Daisy and I watched the colors of dawn dance across the sky. Around us, individual crystals of snow caught the light, glittering with rainbows as the first sun rays crested the mountain. Daisy's teats kept my hands warm while I milked. If the morning was particularly cold, I slid my hand between the side of her udder and the inside of her leg, and her body heat enveloped my hand like a mitten. After milking,

I brushed Daisy's body, attending to every inch. At dusk, I layered up again, sat on the rock, leaned lightly against Daisy, watched the mourning doves and sparrows fly in for the night, and milked while the masterpiece of the sunset engulfed the western sky and faded to black.

I wanted Daisy to have a calf to raise—both to help me with her ample milk supply and because Daisy was such a doting mother. But calving season was months away, and with it, the opportunity to buy an orphaned calf. After wresting with this dilemma for a few days, I remembered the dairy where I got Daisy. I remembered they had calved year-round to have a consistent supply of milk to sell. For the past few years, I had bought fresh, raw milk directly from a small dairy farmer in a neighboring town whenever I wasn't milking Daisy. He had over a dozen dairy cows, milked them all by hand, and delivered milk around the area. Wyoming had passed trailblazing legislation called the Food Freedom Act, which lifted a number of restrictions on food sovereignty and legalized the sale of raw milk between dairy farmers and their customers. I didn't know if my milkman staggered his calving throughout the year, but I called him on the remote chance that he might have a young calf I could buy for Daisy to adopt.

He said he did have a calf. She was a Brown Swiss heifer calf, just ten days old, but he didn't want to sell her as he was planning to keep her for his herd. He called me back an hour later to tell me he'd changed his mind. His asking price matched his reluctance, but I agreed enthusiastically. The following week, once I was sure the medication Daisy had been given was out of her system, we met halfway between our towns for the handoff. He opened the tailgate of his SUV to reveal a calf sitting alertly on a quilt, her legs tucked beneath her, her coat a shimmering taupe that lightened to cream on her legs and belly. She blinked at us, calm and curious, accustomed to people since she'd been bottle-fed her mother's milk. He carried her to the back seat of my truck and wished us well with the strongest handshake I have ever received. During the ride home, the calf nuzzled my shoulder and neck from the back seat, then curled up and looked out the window as I told her all about Daisy.

I wasn't sure how Daisy would react to the calf. I was prepared for some confusion, some resistance; for the calf to try nursing and for Daisy to walk away. I heaved the calf out of the truck and, standing behind her with a hand on each side of her shoulders, guided her to the front yard. Daisy was at the far end of the yard. She watched me lazily as I shuffled from the truck, stooped over the gangly calf who was obscured from Daisy's view by the rail fence. As soon as I nudged the calf through the gate and into the yard, Daisy stiffened, then galloped over to us with a mama moo that said, "There you are! I've been wondering where you were!" Daisy started licking the calf, who was delighted to receive the attention. Within minutes, the calf nuzzled against Daisy's side and began nursing.

I named the calf Mara, short for *Marasmius oreades*, the Latin name for the fairy ring mushroom—a nod to her mushroom-brown coat, her sprightly nature, and the magic that absolutely played a part in bringing her to us. Though their love match was instantaneous, I wanted to keep Daisy and Mara separate from the other cows while their bond continued to strengthen. A few years earlier, Mike had built a new barn near the house. Half the barn was occupied by other cows who needed special care close to home. I thought about putting Daisy and Mara in the other half, but that side didn't have an adjoining corral outside. We were in the middle of the coldest winter we'd had in years, with temperatures hovering between -20°F and 20°F for much of January and February. Being stuck inside the barn during daylight hours would have been colder than being outside where they could soak up sun rays with their fuzzy coats. I could have taken Daisy and Mara down to the old barn at the corral, but I wanted them close. And so I kept Daisy and Mara in the front yard. They were perfectly content in the yard, close to the rest of the herd; and with Daisy in the yard, I wouldn't have to lug full milk pails across the treacherous sheet of ice that covered our driveway. But if Daisy and Mara were going to stay in the yard during blizzard season, they needed some sort of shelter. Mike and I rearranged a few fence rails, and Mike backed the horse trailer to the north edge of the yard. The open trailer had full southern exposure, and I filled it with a bedding of straw. Voilà, a portable barn. I mucked out the trailer daily and added fresh straw each evening. Daisy and Mara spent their days sunning themselves in the yard and slept in the trailer at night, curled up side by side in the straw.

The hush of deep winter brings a certain intimacy. When flakes fill the air, the apparent world shrinks. The senses are quieted in this stillness, in the reverent silence. It's a swaddling of the soul, akin to what I imagine the devout experience in an empty church. Milking Daisy that winter was unexpectedly tender. Daisy was more patient with me than she'd ever been in the past. I wondered if, after her miscarriage and before we got Mara, Daisy had accepted me as a sort of surrogate calf. She was as gentle with me as she had always been toward her calves, to a degree she had never before reached with me. Our bond deepened. Our trust in each other deepened. That winter, she never showed impatience with me, never tap-danced if my milking took longer than she wanted it to. She stood like a statue for me. Sometimes she ate while I milked, but often, she meditated as she did when nursing a calf, her eyes half closed, her posture relaxed. Sometimes she fell asleep. As I sat beneath her with my head and shoulders resting against her warm belly, rocked gently by her breath, there were times I almost fell asleep, too.

Sometimes I paused during milking and leaned against her side and sipped a mugful of her warm, rich milk, frothy from the velocity of milking. When I worked an espresso stand as a teenager and had maxed out on triple shots, I'd make myself steamed milk with a spoonful of almond syrup stirred in. That's what Daisy's milk tasted like, sipped in the snow, milked into a mug and enjoyed immediately. When I finished milking, before I brushed her in gratitude, Daisy groomed me. With a gentle toss of her head, she covered my body in long, deliberate swipes of her tongue. I only ever let her groom my clothes, for cow tongues are rough enough to take off a layer of skin with one lick. Sometimes, I misjudged the length of her tongue and she nicked my cheek with her spiny taste buds and I flinched in pain, but it was worth it, to be so loved by Daisy.

I cherished those hours bookending my days, the ritual work of milking, my body and mind totally engaged with Daisy, the whole of the uncompromising natural world around us. Milking a cow by hand in the heart of winter was objectively harder than attending to the administrative tasks of my business and my life. Physically, it was far more uncomfortable than sitting in a soft chair in my heated office answering email. And yet, even in temperatures so cold that hoarfrost coated Daisy's eyelashes, winter milking left me feeling as bright as sunlight glinting off snow. While I was milking, and after I was done, I felt restored. It left me fortified in a way office work and email never have. It left me feeling *right*.

I developed a theory to explain this seeming contradiction while reading Michael Pollan's *How to Change Your Mind*, which is about psychedelics. The chapter on neuroscience introduced me to a network of regions in the brain called the default mode network. This central hub has been dubbed the home of the ego, the orchestra conductor of the brain. When the default mode network is active, other networks of the brain communicate primarily to and through the default mode network, which filters and organizes the extraordinary chaos of this information. The default mode network controls our narrative of who we are, how we think about ourselves, and our interpretation of our identity. It ignites and defines our sense of self. It grants us our drive and our ability to reason. It's where we time-travel—projecting into the past and the future—and have the imaginary conversations we're excited about or stressing over. It's active when we worry, when we judge, when we reflect on ourselves.

Multiple studies have shown an overactive default mode network to be associated with depression, anxiety, addiction, and overall unhappiness. We can become trapped by this part of the brain, stuck in deep ruts of habit and rigid thinking which this network can generate and self-perpetuate. Even if those outcomes are avoided, inherent to a sense of self is a sense of separation from all that is not "self." The default mode network can create an imagined hierarchy of self—the impression of being more important than anything and everything else. We can easily become the center of our own universe. We can become seduced by the illusion of borders, of duality, of us versus them, of me against you. It would seem an intensely active default mode network doesn't just self-perpetuate these more harmful traits within an individual brain, but that, unchecked, they can infiltrate a society.

While researching the effects of psychedelics—which are proving to be a notable treatment for depression and addiction—English scientist Robin Carhart-Harris took fMRI scans of the brains of volunteers after they had been given psilocybin. An ocean away, at Yale, researcher Judson Brewer saw the published images of these fMRI scans and noted how strikingly similar they looked to the fMRI scans he had made of the brains of experienced meditators during meditation. Both sets of scans showed substantially decreased activity in the default mode network. When the default mode network is not engaged, other regions of the brain communicate directly with one another in an ornate web of connections. With this shift comes heightened creativity and perception, flexible thinking, and out-of-the-box problem-solving. Time-traveling is traded for being fully present in the moment; judgment is replaced by wonder. Our sense of self recedes. Concepts of "I," "me," and "mine" fade in favor of heightened unity—a feeling of being connected to others, to the Earth, and to all that is. As Carhart-Harris states, "Ego dissolution and [a] sense of oneness are essentially synonymous. You can't have one without the other."

Across time and traditions, meditation is associated with separating from the reflexive thoughts that seem to enter our minds of their own accord and carry us away with them. Many meditation techniques focus on mindfulness, focused concentration, or repetition—of a mantra, of cycles of breath—as a means of overriding or detaching from these thoughts. When I was milking Daisy, I entered into a meditative state, guided by the rhythm of my hands and the rhythm of Daisy's breath, woven through with birdsong and sunrise skies. During that time, I was not the center of my thoughts. During that time, I was immersed in the present moment. During that time, my default mode network was disengaged. I thought about the other times I felt this way—when I lost my sense of self through tasks that were acts of mindfulness. Chopping wood is a kind of meditation. Sitting in the sagebrush with binoculars, watching a cow give birth to her calf is a kind of meditation. Lying on Fiona or Sir Baby is a kind of meditation. In fact, for me, these are all more reliable forms of meditation than sitting on a pillow trying to meditate. The physical component demands total presence, leads the way for my brain to disconnect from thoughts of the past and future, of myself, of worry. There's just no room for such thoughts while I'm engaging, wholly and mindfully, in the present moment as these tasks require. Perhaps this is part of the reason why I enjoy spending so much time with animals that could seriously injure me (whether because of their size or their fangs) were I to get distracted while with them.

This led me to wonder about all the work humans have done in order to survive in previous eras that were, by their very nature, tasks of mindfulness, "mini meditations." Knapping an arrowhead. Weaving a basket. Chopping down a tree with an ax. Making fire without matches. Stalking an animal. Navigating by starlight. Did these tasks, these "mini meditations," essentially build in time and space for the default mode network to quiet down, allowing opportunities for creative insight, expansive thought, and a sense of interconnectedness—with others, with the Earth, with the mysteries of life—to regularly infiltrate our ancestors' days and nights? What might the cumulative effect of performing these daily "mini meditations" be on a person's brain? On a society in which these tasks were the norm? Did much of the work of survival across most of human history actually provide balance for our ancestors' brains and well-being?

What happens when we lose those daily "mini meditations"? So much of modern work and modern life has eliminated the tasks that help quiet the default mode network, and has replaced them with tasks and circumstances that keep this network of the brain unrelentingly engaged. The default mode network is active while we're lying in bed stressing about the backlog of email clogging our inboxes, multiplying faster than we can respond. It's running on all cylinders when we're tight with worry about debt or inadequate healthcare, when our mind revisits the trolling we were dealt online, when we move through the world knowing our race or gender or sexuality may be used by others as license to harm us, when we're consumed by pressure to optimize every moment of our time, when we're battling the precariousness that defines and permeates so many aspects of our lives. Behind our burnout, behind our anxiety, is a society that has left no space for the default mode network to release its grip on our brains.

I wonder if having time and space for the default mode network to disengage provides a crucial balance, a kind of neurological restoration, essential in order to be truly healthy. That as we deplete our reserves—the reserves of our spirit, for lack of a better word—we need restorative periods to revitalize and replenish our resources. That perhaps we are as vulnerable to unhindered extraction as the Earth herself—that through the unhindered extraction of our time and attention, we are left without the space for rest, renewal, and wonder that our brains truly need. I'm not a neuroscientist, but I know how I feel when the circumstances of my life leave little room for my daily "mini meditations." I know how drained I became—constantly, inescapably drained—before I created strong boundaries around email and released myself from the guilt I felt about not responding to all the correspondence I received. I know how I feel when chronic worry and stress infiltrate my mind to such a degree that it's the last thing I'm conscious of before I fall asleep and dominates my thoughts the moment I wake.

If we lose time and space for the default mode network to disengage, do we end up depleted, our spirits as barren as denatured soil? Perhaps, as is true in nature and in agriculture, consistent practices that nurture the whole of our being are the most sustainable, have the most lasting positive impacts, and grant us the highest degree of health and productivity. Heath and productivity are synonymous in the natural world. Why wouldn't this be true for us as well? When we have time and opportunity to transcend our sense of self, the mind is free to soar, create, dream, contribute, play. How can we reach our potential without this?

During the times I have allowed myself to unplug completely—a week away from email, social media, and administrative work; on very rare occasions, a month—I always expect to blob around, to use that time and space for little more than healing the burnout I can no longer deny. Instead, I find myself more productive, more creative, happier, and healthier than I'd been in the months leading up to those breaks. My days are full, bursting with activity and physical labor, yet I don't feel like I'm constantly racing against the clock and the calendar. I feel autonomous from the concept of time—time feels like a well into which I can dip to my heart's content. I think back to how I felt during the summer I spent on the mountain, washing my clothes by hand and traveling afoot, almost completely disengaged from the modern world. My digital breaks, when I allow myself to untangle from the tentacles of modern work and modern stress, are the only times I have been able to replicate the way I felt on the mountain. I'm still busy—I'm still out milking my cow and building fences and working with my animals. I've found that it's not the omission of work that prevents and heals my burnout, but the omission of certain types of work—the ubiquity of screens, the tyranny of email, the time on the phone spent sitting on hold with entities to which we are nothing but numbers, the unnatural dichotomy of suspension and frenzy that is traffic, the pulse of the unrelenting grind—work that keeps my default mode network engaged without respite. My happiest days, my most fulfilling days, my most memorable days aren't the ones that are free from work or even free from discomfort. They're the days I'm muddy and windblown, part of something much larger and grander than myself, and both my phone and my default mode network have been dropped somewhere on the path behind me.

CHAPTER 10

The 1% of Cows

Winter eventually retreated. The snow melted and spring unfolded. Roxy, Daisy's calf born two years prior, was close to calving for the first time. I watched her closely along with the other heifers, but since half of Roxy's genetics were dairy, her udder was a less than reliable sign—it just kept growing. We were days away from a new moon, and finding Roxy at midnight or 3:00 a.m. took a bit of luck and a lot of wandering with the weak light of a flashlight, searching for where she was sleeping. When it's as dark as it is without moonlight, you have to stumble right on to a black cow to see her. I had more success finding clusters of sleeping cows by sound rather than sight, listening for their collective deep breaths in the dark, huffs of sighs, their great exhales. Once I found a group, I had to shine my light on each cow to find Roxy before returning to bed. If she decided to wander off alone to find a secluded spot to have her calf, as cows in labor often do, finding her in the dark would have been nearly impossible. And so, when the back wrinkle in her udder filled out, I put Roxy in the new barn near the house so I could check on her more easily.

Since Daisy's miscarriage, I had been putting off going to town until it was absolutely unavoidable. There was too much to take care of at home, and town was thirty miles away, and errands are never something I look forward to, even in the mellowest of times. I had been averaging one trip to town every three or four weeks, and by mid-April, I had reached the point where I could put it off no longer. I coordinated with Mike. I would check on Roxy midmorning, then go to town for a few hours. He would be home by noon to feed cows and would check on her then. I got home around 1:30 p.m. and went straight to the house with groceries. As Mike and I were chatting, I went out on the deck with binoculars to peek at Roxy in the barn. I could barely see her, but something about her posture set off every alarm in my head. "I think she's calving," I said as I ran past Mike and dashed out to the barn. Mike was right behind me. When we got to the barn, we saw Roxy was in labor, her calf's hooves were already out, and her calf was breech.

You can tell if a calf is breech or not by the hooves when they emerge. When a calf is in the ideal position, the shiny black part of their hooves are facing up. These are the front hooves, and the calf is positioned like it is diving out of its mother—front legs, then face, then the rest of its body. When a calf is breech, the soft white under-

sides of its hooves are facing up. The calf is in the same diving position, but backward—the hind legs come out first, and you are looking
at the soles of its hind hooves. We saw the white bottoms of the
hooves and Mike went straight into denial. "Maybe it's just twisted
around and looks backward," he said. Since Roxy was tame and
trusted me, I walked up to her, pet her back, and reached my hand
inside her, following the leg of her calf. "I feel the hock," I said to
Mike. The hock is the pointy joint on the hind legs of cows (and cats
and horses and many other four-legged animals) that does not exist on
the front legs. "I'll call the vet," I said, and we ran back to the house.
The vet's receptionist said he was in the field and wouldn't be available
until five o'clock. Babies don't wait, and this one was coming, so Mike
and I had no choice but to deliver it ourselves.

Delivering a breech calf is stressful because if you do it wrong, the
calf will die. Cattle have relatively short umbilical cords, and when a
calf's abdomen emerges, the umbilical cord breaks, and this compels
the calf to take its first breath. I had often observed the bizarre phenomenon in nonbreech births where a calf's front legs and head were
entirely out of the cow, and sometimes it hung out like this for a while
if the cow needed to rest, and the calf was not breathing, for it was still
receiving oxygen from the umbilical cord. But in breech births, the
abdomen is delivered before the calf's head. If the calf's head is inside
its mother when the umbilical cord breaks—prompting the biological
demand to breathe—the calf will have no access to air, and it will die.

Breech births in cows can be hard to deliver successfully because
the two main hang-ups during delivery are the calf's hips and shoul-

ders. These are the widest parts of the calf, by a significant margin. Depending on the size of the calf and the experience of the cow, she may need time to push these parts out, which is why I've seen the strange scenario of a calf with its front legs and head hanging out for several minutes before the cow delivers the shoulders and the rest. With a breech calf, there can be no waiting. Once the hips are out, the abdomen slithers out quickly because it is so much narrower. The person delivering the calf must get the shoulders and head of the calf out immediately, or the calf will start breathing while its head is still inside the cow. Breech births require human assistance because the cow just can't do this fast enough on her own for the baby to survive.

And it was up to Mike and me. We got the chain and the calf puller, which is a huge T-shaped bar that fits against the hind legs of the cow, with notches in the long part of the T and a ratchet handle. I grabbed leather gloves and a towel, because baby calves are slippery and we would need something with a little traction if we had to pull by hand. Out in the barn, we found Roxy lying down and pushing. I knelt at her side and looped the small, rounded chain around her calf's ankles. A chain is used as the point of contact because the force required to pull a calf in an emergency is so great and calves are so slippery, a chain has the least margin for failure.

Mike hooked the chain to the calf puller. I sat beside Roxy and watched her contractions, telling Mike to ratchet when Roxy pushed, and to stop when she paused to rest. Slowly, Mike and Roxy delivered the calf's hips. And then everything accelerated to warp speed, and at the same time, each millisecond felt drawn out like taffy. When the calf's hips emerged, the abdomen slithered out faster than I thought possible. Mike hit the wall of the barn with the puller bar and didn't have room to ratchet out the shoulders. We each grabbed a calf leg with both hands and pulled with all our might while I shrieked, "We have to get it out *now!*" and somehow, together, we did. I thrust my hand down the calf's throat and scooped out a handful of jelly-like liquid as the calf blinked and gasped and shook his head. We did it. We placed the calf at Roxy's head so she didn't have to get up to lick him clean, and left them together to rest while we returned to the house to purge the adrenaline coursing through our bloodstreams.

When Mike and I returned to the barn an hour later, both Roxy and her calf were up and her calf was trying to nurse. Roxy stood perfectly still while he latched on and had his first meal. Mike and I sat together in the hay, still overwhelmed from the delivery, and watched the simple beauty of Roxy nursing her calf. "How many people get to experience this in their lives?" I wondered to Mike. He didn't know. I didn't know this kind of life existed fifteen years ago.

Because the breech birth was hard on Roxy, and because I wanted to watch her calf closely for any sign of illness or respiratory issues—I wasn't convinced he hadn't breathed any liquid—I kept them in the barn for monitoring. Roxy's udder was large for a heifer, thanks to Daisy's genetics. By evening, after her calf had nursed twice and her udder still looked tight, I decided it would be best for Roxy if I attempted to milk the two teats her calf hadn't touched. I had always frozen Daisy's leftover colostrum for future orphans and twins, but I hadn't been able to save any colostrum from Daisy after her miscarriage.

Roxy had more colostrum than her newborn could drink, so along with helping to relieve her udder, I could replenish our colostrum stash—if Roxy let me.

Roxy had let me rub her udder before she calved. She loved attention, and when I rubbed her udder, she would raise her hind leg to encourage me to scratch that hard-to-reach spot where her inner leg met her belly. With her calf lying in the hay in front of her, I crouched at Roxy's side with my milking cup. I balanced on the balls of my feet, ready to leap away should she kick, and tentatively began milking. Roxy stood calmly, serenely, and chewed her cud. I milked over a half gallon of colostrum from her back teats.

The next morning, Roxy's calf was running circles in the barn, bucking with glee, but Roxy had a bulge of fluid under her belly skin. Her udder was tight, and the texture of her udder was almost like clay. An internet search told me this was udder edema and that it was common in dairy heifers freshening—making milk—for the first time. Roxy fit the profile and all symptoms. The cure was to make sure the cow was completely milked out twice a day. I was grateful that Roxy was so easy to milk. I simply sat down beside her with my pail and she stood for me without food or halter or head catch, requiring nothing but gentleness and conversation.

Two days after Roxy calved, I glanced out the window at a cow walking up the driveway. "That cow is going to calve today," I said to myself, and took a closer look to see who it was so I could keep track of her. I don't know what it is I see when I look at a cow and know she's going to calve—it's something in her posture that I can't explain or point out to anyone. It's the subtle noticing that begets knowing that Gavin de Becker emphasizes in *The Gift of Fear* (though in a completely different context)—information we perceive yet can't articulate, and we call it instinct or intuition. I've learned to trust it. But in this instance, I was wrong. It was Six, Mike's eldest Grandmother, and there was no way she was going to calve because there was no way she was pregnant. She was nineteen, the equivalent of nearly ninety in human years. She was extremely bony, underweight due to her age, and her bones creaked when she walked. I must have caught Six's geriatric hobbling out of the corner of my eye and misinterpreted it.

Mike got home a few hours later and saw Six lying down, off by herself. He was worried that she had wandered off from the herd because she was going to die. He went out to check on her an hour later, turned on his heel, and came to get me. Six had calved. Six, our ancient Grandmother, had given birth to a beautiful, perfect little calf, smaller than average but not by too much. The calf was lively and healthy, already up and prancing around her mother. I still don't understand how it was physically possible for Six to carry a baby in her condition—by all measures it was impossible. She and her calf were truly miraculous.

Six did not have any milk. This was no surprise. She had given all her resources to her calf in utero and had none remaining with which to make milk. But she doted on her calf. She stood and licked her calf's little body. She let her calf suck her empty teats. I made a bottle from Roxy's colostrum and Six's calf gulped it down. She wanted more. I had milked Roxy that morning as part of our twice-daily treatment for udder edema, but I hadn't yet milked her a second time that day. I led Six's calf into the barn with Roxy.

Most cows will kick at any calf who tries to nurse other than their own, but sometimes heifers are less rigid, to the point of it being detrimental—if they allow thieving calves to take their milk, their own calf can suffer. I had the calf bottle with me to milk into if Roxy balked. I pet Roxy and positioned Six's calf beside her. I rubbed Roxy's udder and squeezed her teats like I was going to milk her, then let the calf take over. Roxy stood peacefully. I brushed her while Six's calf drained her udder and had a satisfying meal. When the calf finished, I took her back out to Six who had made her way to the front of the barn. Six licked her calf from nose to hock, and the two of them lay down side by side. Mike put panels up around the front of the barn to make a little corral around them, and I set up a small trough for them and filled it with water. We gave Six a pile of hay. She was going to need help raising her calf, and Roxy and I were going to do it.

The next day, Star Baby calved. Star Baby was Star's first calf—Star was the calf born in the BLM during the yucca frenzy so many years before. Unlike her mother and most of our cows, Star Baby had never really been fond of people. She wasn't fearful, she just wanted nothing to do with us, unless it was winter and we brought her food. If I tried to pet her, she'd toss her head as if to say, "Get your filthy hands off me," and walk away.

When I saw Star Baby calving on a hill at the edge of the pasture, I crept out to make sure her calf was not breech. Even with binoculars and distance, Star Baby saw me and stood, walked a short distance, and lay down again—this time behind a sagebrush, facing me with her backside arranged so I couldn't see her calf's hooves. I walked down the hill and circled behind her. I wasn't as sneaky as I'd hoped, but Star Baby tolerated my presence and her delivery went smoothly. Once her calf shook its head and Star Baby stood to lick it off, I went back to the house and kept an eye on them with binoculars to make sure her calf was able to nurse.

By late afternoon, Star Baby's calf still hadn't had her first meal. She nuzzled around Star Baby's udder and tried to nurse, but she never managed to get a teat in her mouth, even though Star Baby stood for her patiently. The calf could not figure out Star Baby's udder. Star Baby's teats didn't dangle down. They angled forward, perfectly positioned to slide right into a calf's mouth as long as the calf was coming at the teat head-on. Star Baby's calf had been trying to nurse from the side and hadn't managed to draw a teat into her mouth from that angle. She needed to get colostrum before nightfall. I had to intervene.

With many of our cows, I can approach them in the pasture and help maneuver their babies onto teats for the first time, if need be— the cows will stand calmly, more focused on their calf than on me. But Star Baby's disdain for people thwarted such simplicity. When I walked out to Star Baby and her calf, she hustled away, mooing for her baby to follow her. Down the hill she went. The more I followed, the farther she traveled. Mike came out to help me, and together we tried to get Star Baby and her calf into the corral. Star Baby trotted in sweeping figure eights, her udder swinging, her calf prancing after her. A gusty windstorm blew in and red dirt filled the air and blinded us as we chased after Star Baby. We finally got them in the corral, our skin and teeth gritty from dirt, just as the sun was going down. As I latched the gate, Star Baby, furious about being contained, tried to bite me. I had never had a cow try to bite me before. I told her that all we wanted to do was help her and her calf. I always tell the animals my intentions, the "why" behind what I'm doing with them, if they seem stressed.

Sometimes, helping a cow or a calf is not pretty or graceful or easy. Sometimes, a calf will look at me with wild eyes that scream, "You're scaring me, you're hurting me, PLEASE STOP!!!!" and I say, as gently as I can, "I'm trying to save your life." Sometimes, my own life feels just as painful and terrifying, and I've started playing with the idea that perhaps, in those moments, some benevolent powers-that-be— powers unfathomably stronger and wiser than I—are trying to help me out; it's just that from where I am, it feels like an attack.

I lured Star Baby into the head catch with some hay. As she ate, she began to relax, and her breathing calmed and slowed. While she ate, I helped her calf stand in the right spot to get one of Star Baby's torpedo teats in her mouth and nurse. Star Baby's udder had puzzled me in the weeks before she calved. Her left front quarter did not inflate like the rest. I couldn't figure out if she was letting another calf suck that quarter—extremely unlikely with an older cow—or what the explanation might be. Now that I had a chance to study her udder up close, I understood. Star Baby must have had undiagnosed mastitis in that quarter the previous year, and now that quarter was dead—it was not producing milk at all.

The year before, I had been so overwhelmed with work, I wasn't out with the cows as much as usual. Mike is not one to stare at udders the way I'm inclined to do as a milkmaid; and Angus cows are less at risk for mastitis to begin with, because they don't produce surplus milk—their calves keep their udders drained and healthy. But this spring, Star Baby's udder was enormous for an Angus cow. I didn't remember it being so big in the past, but it must have been big enough the previous year that her calf ignored one quarter consistently, and it eventually got mastitis. Luckily for Star Baby's calves present and future, her remaining three quarters held an abundance of milk. So much that her calf was not going to be able to drink it all until she grew bigger and more voracious. Though Star Baby's calf nursed until she was satisfied, she only drank from one quarter. I had to make sure all three quarters were milked out every day or risk Star Baby getting mastitis again.

I had put Six's calf in with Roxy twice that day, but I knew she could drink more. She was getting adequate milk from Roxy, but not enough to properly stuff herself. After Star Baby's calf finished her first meal and lay down, I brought Six's calf to Star Baby's side. Six's calf grabbed a teat with her tongue and drank until she couldn't hold another drop. When she was full enough to burst, she waddled back to Six, who licked her and loved her, though she could not feed her.

Even after both calves had their fill, Star Baby's udder was still full of colostrum. I knew Star Baby's calf would nurse a few more times during the night, but I didn't want to neglect Star Baby's udder, considering her history. I got a pail to milk the remaining teat that neither calf had touched. Star Baby's teats were short and almost horizontal to the ground. I couldn't use my full hand to milk—her teats were not long enough to grasp—and had to squeeze with just my fingertips. Star Baby was relaxed while I milked her, just as Roxy had been. She was calm and patient. Or perhaps she was just pleased that I was kneeling in the dirt at her feet.

Since they all needed help from one another, I kept Roxy, Six, and Star Baby, and all three of their calves, in the barn and adjacent corral. Every morning, at first light, I joined them and helped them help each other. Robins, mourning doves, blackbirds, sparrows, and sandhill cranes chattered and trilled around us, rejoicing in the return of spring. The cows anticipated my arrival. When I let Roxy out of the barn, she trotted to the head catch. She had a haystack in the barn to eat from as she wished, so I knew she wasn't motivated by food— she stood in the head catch just because she wanted to. Six's calf had learned quickly that if a cow was standing in the head catch, that cow was *hers*—her meal, her job, her delight. She galloped over, wriggled alongside Roxy's body, and gulped down whatever milk remained af- ter Roxy's calf had breakfasted. When I let Roxy out of the head catch, Star Baby was lined up and ready to go in. Star Baby had realized that whenever she was in the head catch, she got extra food. I piled hay in front of her, and while she ate, Six's calf gorged on Star Baby's bountiful udder. Star Baby's calf always ate first, but she was still only nursing from one quarter. While Star Baby was occupied, I carried an armload of extra hay to Six, for she was still so skinny.

One morning, Mike joined me during my musical cow routine. "Six's calf has three mothers," I said, laughing, and he said, "We should name her 3M." 3M for three mothers, particularly special because three is half of six. While 3M got her food from two other cows, she got love from Six. After 3M consumed all she could hold from Roxy and Star Baby, she trotted back to Six, who licked and nuzzled her. They stood together. They slept side by side. The bond between them was humbling in its purity. I fell in love with 3M immediately. She understood from day one that I was going to help her get her meals, but that she was going to have to make the most of every opportunity, to eat fast and suck hard. She ran to me when I arrived at the corral, licking her lips in anticipation, and bounced into the alley whenever a cow was in the head catch. She was scrappy, happy, eager, and opportunistic. She was smaller than the other calves, but she was tough and determined.

This morning routine never felt like a chore—it was ritual beauty I got to help orchestrate and take part in. It was a time of guaranteed wonder, of peace, of the warm spring air on my cheeks. Of doing the good work of helping a cow raise her calf and helping two cows keep their udders healthy. Of filling the water tank, of watching the cows and calves interact—the group together, the community of the little corral—each one independent and respectful, familial in the very best sense of that word. We repeated this ritual every evening, and I treasured that time, too. My time at the barn was the best part of my every day in a quiet, dependable way. By evening, the side of the barn where we had the head catch was in shadow, but the earth held the warmth of the day. I sat on the ground as 3M sucked and slurped ecstatically, and I breathed—really breathed—and smiled—really smiled.

By the time Roxy's calf was a week old, he was drinking exponentially more than he had as a newborn and could consume all the milk Roxy produced on his own. Roxy no longer needed 3M's help to keep her udder drained twice a day, and Roxy's udder edema had completely vanished. My early concerns about the health of Roxy's calf had vanished, too. He was healthy and strong, and he and Roxy had outgrown the barn. It was time to let them out.

And so, my beautiful mornings and evenings transformed, became even more intimate. It was now Six and 3M, Star Baby and her calf, and me; though Roxy kept returning to the barn, kept wanting to stand in the head catch just for the fun of it. Star Baby, for all her haughty disdain, was surprisingly generous with 3M. She refused to allow 3M to nurse when they were out in the corral, but as soon as she saw me approaching, Star Baby walked to the head catch and waited for me there. Once she was in the head catch, she stood calmly and patiently for as long as 3M wanted to nurse. I cherished those weeks of mornings and evenings with Six and 3M and Star Baby and her calf. I've been around long enough to know that if you love something, you better revel in it with all you've got, because nothing lasts. Calves grow up. Seasons change. And even when seasons cycle back around, no two springs are ever the same, no two winters. Everything is temporary. These chores were sacred because I knew they wouldn't last. They couldn't last. Nothing does.

Rain. Our single, four-letter word for water falling from the sky is grossly inadequate. There should be more words. Seattle rain and Wyoming rain are as related and dissimilar as chihuahua and wolf. Wyoming rain is the wolf. It can appear, seemingly from out of nowhere, and surprise you dangerously. It can be terrifyingly violent. Despite this, it is revered. Rain brings grass, and grass is life.

When it rains in Wyoming, it's like a giant barrel being dumped out from the heavens. It's like a blizzard in its liquid state. Wyoming rain comes down hard and fast, each raindrop as long and thick as your finger. Where the ground is bare, the clay-laden earth can't absorb such a deluge, and water stands in enormous puddles, runs rapids down the driveway, and the top six inches of the earth becomes sticky, slicky gumbo—clay mud that can eat shoes and is nearly as treacherous as ice. The first time I lost control of my truck was not on ice but in Wyoming mud.

Still, these torrential downpours don't fill water troughs for the cows—that was my job. Calves are hungry no matter the weather, and 3M needed my help with Star Baby. Mike had left, and when I finished my coffee, I put on the ankle-length snow skirt I hadn't touched since deep winter, a rain slicker with a hood, and my Muck boots and slopped out to the corral. My trudging turned to sprinting when I saw Six was down and thrashing. She was lying on her side in the mud, legs splayed, kicking the air. I reached her side and gasped. Her head had slid beneath the bottom rail of one of the corral panels and she was stuck. She was writhing and flailing and frothing at the mouth, trying to rescue herself. I splashed to the barn and grabbed a lariat—a stiff lasso rope—and threaded the loop around Six's neck. I pushed it as far down as I could, down to the thick base of her neck where it met her chest and shoulders. I looped the loose end of the lariat around a rail of the metal barn gate for a makeshift pulley, then heaved with all my might. I managed to slide Six far enough away from the panel to free her head. The slipperiness of the mud was the only reason I succeeded—even old and frail, she still weighed in at close to a thousand pounds.

Six immediately began struggling to stand. When a cow gets up, she rocks a little for momentum to launch herself up and onto her hooves, similar to the movement we would make when getting out of an oversize plush chair. Cows can kill themselves trying to get up when they're stuck. If they're mired in mud, or up against a wall without enough room to rock back and forth onto their feet, they will keep trying until they die of exhaustion. Crouched next to Six, I noticed the earth sloped gently away from the wall of the barn in a slight downhill grade—unremarkable on a dry day. But in the rain, that slope had been perilous. I imagined that Six must have laid down under the awning of the barn—usually a dry place, but on such a day, with sideways rain, not dry at all. I imagined that when she tried to stand, she wasn't able to rock all the way up to standing because of the slippery mud. Instead, the momentum of that movement must have caused her to slide ever so slightly down the tiny slope. I imagined that this positioned her body and head downhill from her legs and hooves. To get up from such a position, she would have had to fight against gravity and launch herself uphill. And in the mud, with no traction, at her age, she couldn't do it. Six was already weak because she was so old—for the past few days, she had been having some trouble rocking up to standing even when the ground was dry and flat. And so, she must have struggled, and as she struggled, she slid closer to the corral panel, and when she tired, she lay her head on the ground. I imagine this was how she ended up with her head under the railing.

I took the lariat from Six's neck and tried to help her sit up, but she was too weak and exhausted to hold her head up. She thrashed in the mud, trying to get up, too weak to succeed. I caught a glimpse of a raw patch of skin on her shoulder where the hair had been worn off during her struggles. Desperate to keep her from thrashing, I sat beside her head with my hands on her face. As long as I was touching her, she lay still. I didn't dare splash back to the house to get my phone to call Mike—when I moved away from her, Six started convulsing in attempts to get up, and I didn't know how much longer she could fight before expiring. I wasn't willing to risk it. So I stayed with her in the mud in the pouring rain, waiting for Mike to return, not knowing when that might be. "Don't die," I sobbed through the downpour. "Not now. Not today. Not in the rain. Don't die yet, don't do it, please." Her eye, the one I could see, was wild. It bulged out, then sank into her skull, and I was sure she was going to perish in the storm as I pet her cheek.

When Mike got home, he found us in the mud and quickly rearranged a few of the steel panels between the corral and my garden. Once Mike opened a passageway, he grabbed one of Six's hind legs while I pushed her shoulders, and together we slid Six out of the corral and into my garden. We maneuvered her to a level grassy patch and got her sitting up—lying down but with her head up. She seemed relieved. The only question was whether Six could get up at all, whether she would ever get up again.

After taking care of 3M and Star Baby, who had been huddled in the rain, bedraggled but fine, Mike and I returned to the house. I escaped to a hot shower. The rain traveled past and the sun came out. I nervously peeked out the door at Six in my garden, and she was up! She was standing, peacefully grazing the luscious, rogue grass that was already six inches tall between my raised beds. She moved delicately enough that I wasn't worried about her destroying my garden the way another cow would. She nipped down all the grass and didn't even touch my raspberries.

After she grazed my garden, and after the corral dried out, we moved Six back to the corral because we were afraid she might lie down between the raised beds and get stuck. We wanted her to be comfortable and safe and close to 3M. Sitting in the mud and the rain with Six, I had begged her, "Don't die yet." Yet. I knew it was coming. We all knew it was coming. She created her perfect calf and delivered her calf safely into the world, and then Six started declining. The day after she calved, it was as if she took a deep breath, thought to herself, *I did it*, and let go. She got visibly thinner daily, and she was already so thin. She got visibly weaker daily, and she was already so weak. By the end of April, when she started having trouble getting up, even when rainstorms and mud were not part of the equation, there was no denying that it was just a matter of time.

When Six could no longer get up at all, we carried water to her in a five-gallon bucket. We brought her hay and laid it on the ground in front of her. We gave her everything we could—water, food, assistance, peace, gratitude. 3M slept beside her and stood at her head while Six licked her from where she lay. Content and carefree, 3M scampered back to Six after getting her meals from Star Baby, and Six, stoic and doting, licked her slowly and methodically from chin to tail as she'd done since their first hour together.

She died in the night. A warm, calm night awash with moonlight. Mike took her body away at dawn, and when I went out to do my morning chores with the corral crew, Six was gone. I put Star Baby in the head catch and fed her treats while 3M nursed, and Six wasn't there, and she wouldn't be there ever again. The corral seemed so much emptier without her presence. Death changes everything with its dark alchemy. 3M finished her morning meal and Six wasn't there to lick her as she always had. I grabbed a curry comb and brushed 3M's little body—her forehead, her cheeks, her neck, her chest, her back, her sides, her flanks, her belly. With her eyes half closed, 3M stood and leaned against me as I brushed her. From that morning on, I brushed 3M after every meal.

Six had been granted independence and safety and respect and care from the first day of her life to the last. Her life was unequivocally a great one. And hers was the best death any of us could hope for—peacefully at home, surrounded by kin, knowing her baby was going to be cared for after she was gone. How many cattle die of old age? I can tell you it's way, way less than 1 percent. I joke with Mike that we work so hard every day for so little money, but our cows are the 1% of cows.

Everything is temporary. I knew this. Six was dying from the day she gave birth. I knew this. 3M was born on April 13 and Six died on May 3. Twenty days. Those magic mornings and evenings lasted for twenty days. The motley cow crew in the little corral were together for twenty days. 3M was twenty days old when she lost her mother. It was only twenty days. I was in awe of the time we had together, that little pocket of time and togetherness and cooperation and appreciation. The grace of that time. The grace of Six.

A week went by. A week of mornings and evenings that had lost their sparkle, that felt somber and empty. Star Baby no longer had adult cow companionship in the little corral. Now orphaned, 3M was often curled up by herself. Star Baby's baby figured out how to get out of the corral and left whenever she wanted to explore or play with the other calves, returning only for meals. She was also getting big enough to empty Star Baby's udder by herself, and there was less and less milk for 3M each passing day. I found where Star Baby's baby was escaping and purposefully locked her out of the corral, blocking her return until Star Baby's udder refilled and 3M had gotten first dibs. But when her calf showed up to eat and couldn't get in the corral, Star Baby stood right up against the panels so her calf could nurse through the railings. As patient and generous as Star Baby had been for the past month, her loyalty was to her calf and it always would be. From the start, I knew Star Baby would be a temporary wet nurse and that eventually her own calf would get big enough and greedy enough to take all the milk she produced. This arrangement was reaching its expiration date.

Fiona was the only cow left to calve. A few days after Six died, I was brushing Fiona and felt her calf moving inside her. A protruding lump rose in a gentle yet startling wave under my hand and disappeared. It was thrilling, but the only other time I'd felt a calf move in utero was with Roxy, and so even though my sample size was exactly one, I became paranoid that Fiona's calf would also be breech. When I saw Fiona standing off by herself with that unexplainable stance that told me she was about to calve, I lured her into the front yard so I could watch her more easily. Fiona went into labor a few hours later, and when her water broke, I collected as much of her jelly-like amniotic fluid as I could and put it in the fridge. I had plans for it. I was overjoyed when her calf's hooves emerged and showed the calf was not breech, and I sat in the yard to watch the birth. Fiona, once again, lay down beside me and delivered her calf right in my lap.

Fiona and her calf spent the rest of the day and that night together in the yard. I needed to be certain that they bonded completely. The next day, I put a tiny calf halter on 3M and threw open the gates of the little corral so Star Baby could rejoin the herd and her wandering calf. Before she sauntered off, Star Baby approached me and licked my elbow. It was the most affection she'd ever shown a human. Then off she went, and I led 3M to the yard to join Fiona and her newborn. While 3M was a month older than Fiona's calf, they were exactly the same size since 3M had been so small when she was born. I got Fiona's amniotic fluid from the fridge and poured it all over 3M and rubbed it into her coat. Cows know their calves by sight and sound, but a key identifier is scent, and I thought perhaps if 3M smelled like Fiona, Fiona would adopt her right away. As Daisy's daughter, Fiona made a lot of milk. Not enough that I ever milked her, other than for leftover colostrum, but I had noticed her production increased incrementally with each passing year. The year before, her calf had managed to keep up with the volume of milk she produced, but barely. If the pattern continued, I knew she would have enough milk to raise two calves.

Fiona was a remarkably gentle, sweet-tempered cow. She trusted me, and I trusted her. I crossed to Fiona and, with 3M in my line of sight, stood at Fiona's flank and stroked the side of her udder. 3M and I had developed some simple sign language and she galloped over. She knew what I was telling her. 3M reached out with a tentative tongue, but the moment she grasped a teat, Fiona turned to inspect the calf by her side, balked, and lightly kicked 3M away. Fiona was not fooled by the amniotic goo bath. Then I remembered the trick Mike had stumbled upon years before with the heifer who had refused to let her calf nurse. I took off my coat, placed it gently over Fiona's face, and tied the arms loosely under her chin. I motioned to 3M while rubbing Fiona's udder. She cautiously returned to Fiona's side and I stepped back and stood at Fiona's shoulder. Fiona's nose remained exposed under the hem of my coat. When 3M eagerly latched onto a teat and began to suck, Fiona swung her head to the side to investigate, and her nose found me. She sniffed my torso then, satisfied, began chewing her cud, a sure sign she was relaxed. 3M feasted on Fiona's decadent supply of milk until she could hold no more.

Fiona had always doted on calves beyond her own, licking them as their mothers did, and she often napped with two or three other calves nestled up against her warm, strong body. I didn't know if Fiona would ever adopt 3M as a second calf, but I hoped that introducing 3M the day after Fiona calved would increase the likelihood of a full adoption. Still, there was a solid chance Fiona would choose not to, and I wanted 3M to learn how to get her own meals from Fiona without my help, in case Fiona never fully accepted her.

I kept Fiona, her calf, and 3M in the yard and set up a playpen of sorts, using two panels and the curve of the wooden fence. I put Fiona's calf in the playpen at night so he wouldn't nurse before I woke up. In the morning, I put the blindfold on Fiona and let her calf out of the playpen. He ran to Fiona, and as soon as he latched on, I coaxed 3M to Fiona's other side. Then I stood at Fiona's shoulder, blocking 3M from Fiona's nose. Fiona sniffed her calf, sniffed me, and stood calmly with my coat on her head while both calves nursed. I repeated this in the evening—putting Fiona's calf in the playpen for a few hours in the afternoon and coordinating a mutual mealtime with 3M at dusk.

The next day, when I let Fiona's calf out of the playpen, both calves ran to Fiona and latched on—one on each side. Soon, 3M had not only caught on, but started manipulating the situation in her favor. She was strategic and clever. She figured out that she could sneak onto a teat whenever Fiona's calf nursed throughout the day—she didn't need me around at all, and she didn't need Fiona to be blindfolded. If Fiona balked, 3M moved to the same side as Fiona's calf, with Fiona's calf positioned between herself and Fiona's head, so that Fiona couldn't really tell she was there. After about a week of witnessing 3M's midday cleverness, I happened to see both calves run to Fiona and begin nursing, one on each side of Fiona's body. Fiona swung her head in a slow arc and sniffed her calf, swung her head to the other side and sniffed 3M, then straightened her head to the front and began chewing her cud. In the days that followed, I often found all three together, both calves nursing, Fiona standing peacefully. When I went outside one afternoon and saw Fiona licking 3M, just as she did with her own calf, just as Six once did, I knew they were a family.

They no longer needed to stay in the yard, and they no longer needed me. Fiona's calf and 3M nursed together, grew together, bonded together, raced around bucking and playing together. An outsider who didn't know the whole story would have thought Fiona had twins. But the truth was so much more complex, shaped by synchron-icity and miracles, tenacity and care. I'm reluctant to call it a happy ending because that seems dismissive of and disrespectful to Six; plus, I don't think life is about happy endings. I think it's about joyful, wondrous moments in between devastating moments in between tedious moments in between scary moments in between thrilling moments and, if we're lucky, a whole lot of love swirled into all of it. In such a spiral, where is the ending?

CHAPTER 11

Meditations with Cows

After Daisy's miscarriage, after her trouble conceiving following Roxy's birth, considering her age and the physical toll taken by producing such large quantities of milk, I decided it was best for Daisy to keep all her resources to herself and not get pregnant again. This was not as simple as it sounds—if Daisy had her way, she'd get herself bred. When she cycles, she leaves the herd and strides purposefully to the far fence line nearest the bull pasture. She stands at the fence, staring at Sir Baby and his bull buddies across the lane, mooing urgently for them. Once they catch her scent, they pile against their fence, bellowing back to Daisy, shoving each other aside with their massive foreheads.

I track her cycles so I know when this drama is coming. When the day I've marked on the calendar arrives (or one close to it; her cycle is predictable but not precise) and I don't see her from the house, I head out with a pail of alfalfa cubes, Daisy's favorite treats. I inevitably find her in the far corner of the pasture, wooing the boys. My goal is to lure her to the corral. She's so close to the bulls and is hesitant to leave them, but she doesn't want to miss out on treats either, and so she tentatively begins to follow me. The bulls follow her, walking parallel with Daisy along their fence line. When Daisy realizes they're coming along, she trots after me, confident she will have her cake and eat it, too. I start running, to keep ahead of Daisy, and she jogs after me, and the bulls break into a gallop, a mass of muscle and dust on their side of the lane. When Daisy and I reach the corral, I dump the treats in the feed bunk and lock her up for two days, just to be safe. The tall wooden rails of the corral are a more intimidating barrier for the bulls than a few strings of barbwire.

Daisy is always offended by her confinement, even though the other cows eventually go looking for her, congregate at the railings, and sleep next to the corral just to be near her. I brush her each day she's locked up and give her extra hay, but her expression is always that of the betrayed. This might be the one instance where Daisy does not get her way, where there is not even compromise between us, where I override her will and her wishes. (This, and coming in the house. I have been so tempted to let her glide through the doorway when she tries to follow me inside, but I haven't dared.) I weaned

Mara when she was eight months old and stopped milking at the same time so Daisy could dry off in time to put on weight before winter arrived. It marked the end of an era. I loved milking Daisy, especially that last, unplanned winter I spent at her side. I have loved all her calves. But Daisy's health, her well-being, her stamina, her longevity take priority over what I might want in the short term.

Mara was a delight. She would become my milk cow when she had her first calf. I started working with Mara when she was about six months old, introducing her to a halter, incentivizing her to follow me—or at least the brush. I sat at her side and mime-milked her tiny virgin teats to get her comfortable with my closeness. When I pretended to milk her, letting her teats slip rhythmically through my fingers, she stood completely still, even when we were out in the pasture among the other cows, and swung her head around to lick my shoulder as if I were her calf.

Mara reminded me so much of Frisco. Her lanky build and crescent-moon horns mirrored his, but the resemblance was more than physical. Her demeanor was as joyous and carefree as Frisco's had been. A certain light shone from both Frisco's and Mara's eyes that seemed to broadcast a reminder that we are here, at least in part, to revel in the very act of being alive.

Sometimes, I hallucinate Frisco standing with the other cows. I can tell them all apart, but sometimes, when the light from the setting sun angles across hides and hip bones, for a flash, I see him. And then I remember it can't be. In the years that have passed since Frisco died, I have been present for many other deaths—horses and humans and very young calves and very old cows. Through traversing my grief over losing Frisco, I learned how to be Death's midwife for the others. I learned how to hold that sacred space for them, how to be strong and soft and support them on their way out, and how to tend to myself later. It's never easy. But I know how to do this now.

The circumstances of Frisco's death—of not being with him when he died, of folding my laundry instead—tormented me for years. Two years went by before I could fold laundry without guilt settling in my lungs like hot tar, and another before I forgave myself. It was sunset. The full moon shimmered like a pearl in the pink sky. I was sitting in the sagebrush beside a Grandmother cow who was down. She hadn't stood up all day. I'd gotten a feeling from her a week before that she wasn't going to be alive much longer. As I sat with her that evening, I knew it would be her last sunset. As she was dying, it became suddenly clear to me that even if I'd been at the barn with Frisco, I wouldn't have been there for him when he died. I would have been too consumed with my own pain. I couldn't have supported him and held a peaceful, loving space for him as he died. I couldn't have watched his head sink to the ground and flop without grabbing it and shrieking for him to hold on, to come back. I would have been loud and deranged.

I would have made things worse. If I had been present for Frisco's death, I would have made it all about me.

Witnessing death still delivers a sharp, hard pain to my chest, like getting whacked in the heart with a golf club. I am sad, angry, and confused every time. But now, I'm not so selfish about it. As the sky grew dark and the moon brighter, I thanked our Grandmother cow for showing me that what I couldn't give to Frisco, I could give to her.

As I midwifed death, I midwifed birth. When Fiona's first heifer calf—the first calf she had delivered in my lap—was about to have her first calf, I spent most of the day squatting in the damp wind with binoculars, watching her labor. She had been hovering around another cow's calf, licking the calf and mooing softly to it as if it were hers. She was ready for her own to arrive, and I knew she would be a doting mother. The early stages of her labor had lasted for hours— after she left the other calf, she stood well away from the herd with her tail cocked and licked her sides, but she didn't lie down to push. She shifted her position often, walked in circles, lay down and stood again, obviously working to help her calf enter the world. Her labor was on the verge of taking too long to be normal, too long to be safe. I had my phone shoved down the shaft of my Muck boot in case I needed backup, but I didn't want to stress the heifer by intervening before I knew it was necessary.

She eventually lay down and began pushing with pronounced

effort. A large silver-blue translucent globe emerged from her body. Through the binoculars, I could see the shadowy forms of her calf's two legs within the globe, but I couldn't determine their position and whether or not the calf was breech. I crept closer. I knew the heifer didn't want me near her, so I shuffled on my elbows and knees, angling until sunlight filtered through the orb rather than reflecting off its curves. By the silhouette of the hooves within, I could see her calf was not breech. I relaxed, and marveled at the birth progressing in front of me. The heifer's water hadn't broke and her calf had not burst through the amniotic sac. The calf was being delivered while still fully enclosed in the fluid-filled sac. In humans, this is called an "en caul" birth and is quite rare.

This phenomenon was likely the reason behind the heifer's lengthy labor—a giant balloon, crossing the threshold of the cervix, is not nearly as streamlined as a slick, floppy body. Once the enclosed hooves emerged, her labor progressed with increasing speed. She stretched out on the ground and began pushing in an even, steady rhythm. The orb grew larger and when the calf's head and shoulders appeared within it, I dropped my binoculars in the dirt and sprinted over to the heifer. I needed to rip open the amniotic sac before the umbilical cord broke and the calf took its first breath. I was worried I wouldn't be able to break through the membrane. I was worried I'd have to use my teeth. I was worried the heifer would be startled by my sudden presence, that she would stand and trot away with her baby dangling, suffocating, before I could rip it open. But the heifer was too engrossed in her labor to be bothered by me. I grabbed the amniotic sac with both hands and my fingernails tore through the membrane. I cleared the fluid and wisps of membrane from the calf's nose and mouth as the rib cage slithered out. Eyes wide, the calf sucked in a great first breath as the heifer, with a final push, delivered the hips and hind legs. The heifer lurched to her feet and I bounced away, leaving her in peace with her new calf. When I stooped to pick up the binoculars and glanced back at the pair—the heifer diligently licking off her baby, mooing a mama's melody—it dawned on me that the calf was Daisy's great-grandcalf.

Daisy and I have been together for over ten years. Intimately together—cheek to belly, hand to teat. In that time, she has given me so much more than just her milk. She holds me when I collapse against her when I am sad or scared. Simply breathing in her scent soothes my nervous system when I am angry or anxious. She can make me laugh with just an expression, and her coat soaks up my tears when I cry. Daisy has shown me an example of how to move through the world with grace and determination. She knows what she wants and expresses those desires clearly, unflinchingly. She is generous and nurturing, her care and attention a cocoon of comfort. While she embodies soft warmth, she wields power that can bend steel when she deems the effort worthwhile. She guides and protects all who need guidance and protection, regardless of whether they are related by blood, and she seems to delight in the act—in being capable of providing others what they need—rather than motivated by what she might get in return. She accepts demonstrations of love as gracefully as she gives them.

I got Daisy because I wanted something from her. On the way to the dairy the day we met, I was thinking of nothing more than my desire to get a cow in order to have her milk. As my life intertwined with Daisy's, I learned that the richness of this earthly existence expands when I shift my thinking from "What can I get?" to "What can I give?" Somewhere along the way, Daisy taught me to evolve my thinking even further; showed me that the most powerful, enlivening question I can ask is, "What can we achieve together?" I've tried to apply this to everything in my life beyond Daisy, though it is impossible, at this point, to untangle what parts of my life are separate from her. She imbues every aspect of my life, because knowing her has changed me to my core.

Sometimes, I fantasize about living twenty thousand years ago, which is less about bemoaning the twenty-first century or romanticizing the past than it is about curiosity, and something much deeper than curiosity—of being so in love with this planet that I am filled with longing to experience her beauty and power stripped free from the ubiquity of industrialized human interference. To gaze at a mountain without cell towers interrupting the splendor; to swim in clear water without plastic bags floating by; to hold my breath as wild herds thunder past, unhindered by roads or fences; to walk outside and not hear the engine noise of planes, cars, semis, and lawnmowers; to see the stars in the total darkness of a new-moon night, without any light pollution planet-wide. And yet, if I had been born twenty thousand years ago exactly as I am, I wouldn't have seen the stars at all. I'm too nearsighted to see much beyond my nose without the help of more modern technology.

How do we find the balance between the technologies that help us live and those that destroy the world around us? I spend much more of my time with nonhuman animals than I spend with other humans, and in spite of this—a cynic would say because of it—I hold immense optimism for our species. We, each of us, carry such creativity, ingenuity, tenacity, power. We each demonstrate these traits in such varied ways. Which is a good thing. Diversity is key to survival, to success. And I refuse to believe that we, with our vast, innovative human brains, are incapable of pivoting to a harmonious solution in the face of our social and environmental crises when it is undeniable that the status quo is failing. Failing us; failing each other; failing this planet; failing the exquisite, ecstatic experience of life. I just wonder if we can remember how to trust each other in time. Our ancestors living twenty thousand years ago understood that cooperation was required for survival. Cooperation among people, and cooperation between people and their surrounding environment. We cannot forget this is still the case.

The more we try to separate ourselves from nature—which is impossible to begin with; our bodies are as much a part of nature as the bodies of butterflies and bodies of water—the more destructive we seem to become. Projections and predictions for the future of food almost always feature lab-grown meat. Also known as synthetic meat or in vitro meat, it is animal-free but for starter cells, and heralded as the ultimate environmentally sound answer to our protein needs. I'd love to see an accounting of the fossil fuels required to create a meal of lab meat. The manufacture and transport of the raw materials used; the energy required for climate control of the building and laboratories; the energy used during production; the commutes of the lab technicians; the management of waste. If the true goal is nutritious food produced sustainably with positive environmental impacts, requiring the least amount of fossil fuels, the solution already exists, courtesy of Mother Nature, in the form of animals raised exclusively on pasture.

I calculated the amount of fossil fuels used during the entire life cycle of my grass-finished beef, from the date the steer was conceived to the consumption of his meat. I included the fuel used to cut, rake, and bale the hay eaten in winter by the mother cow during gestation and, later, by the steer himself. The drive to the abattoir. The commutes of the employees on the days the steer transitioned and was butchered. The electricity used during dry aging and butchery. The energy used to run a chest freezer for a year. The grand total for a year's supply of my grass-finished beef is equivalent to the amount of fossil fuel used during seven days of one average American's commute (which, as of this writing, is sixteen miles each way at 22 mpg).

The benefits of large-scale regenerative agriculture are multifaceted: the restoration of native prairie and the exquisite biodiversity that thrives when grasslands thrive; the support of pollinator populations and reduced pollution when polycultures and organic farming practices are employed in place of industrialized monocropping; the abundant production of healthy, nutrient-dense food. And, at its foundation, the nurturing of the health of our soil. Sequestering carbon in the soil through regenerative agricultural practices—and supporting these practices at every scale, in every region, through proactive policy—is one strategy we can implement immediately in our quest to reduce our net carbon emissions and prevent runaway climate change.

How can it be easier to contemplate releasing a sulfur barrier into the atmosphere to block radiation from the sun in order to counteract the heating of the Earth caused by excess atmospheric carbon? Even though a known side effect is that it would turn our beautiful blue sky to a bleached white during the day and a shocking red at sunset? Even though it is unclear how plants would respond? How can it be easier to bring back the woolly mammoth, as genetic scientists are currently attempting, in hopes that Mammoths 2.0 may help sequester the greenhouse gas stored in the Siberian permafrost, which is at risk of being released into the atmosphere as warming temperatures cause the permafrost to thaw? How are these tactics getting more traction, more consideration, more funding than the readily available solution of working with the animals with whom we already coexist to sequester atmospheric carbon safely in the soil? Is it because those other tactics can be patented? Maybe I am a cynic, after all.

Lab-grown meat—all synthetic food, for that matter, and there is much in development—removes our autonomy from one of our most basic and essential needs. Anyone can raise a chicken, even without training, even in an apartment. While apartment-raised chickens would be less than ideal for the chickens and the neighbors, it can be done. How many of us can synthesize lab meat in our apartments?

Any large-scale transition to synthetic proteins will put even more of the food supply, and therefore more of the power, into the hands of the few. So much power and control over our food supply is already consolidated. Three companies control the majority of seeds on the market. The Big Four don't just dominate the meat industry; these transnational corporations have subsidiary companies across a broad range of sectors: synthetic fertilizers and pesticides, soybean processing plants, grain mills, palm oil plantations and refineries, trucking and shipping fleets, and processed foods (including highly processed "meat alternatives"). Most of the packaged food in major grocery stores and gas stations can be traced to just a handful of corporations when you go past the label to find the parent company. Farm and land ownership is rapidly consolidating into fewer and fewer—almost exclusively white—hands. In California in 2017, 45 percent of the total harvested cropland in the state was held by 2.8 percent of California farm owners (and these landowners are not what we visualize when we think of farmers—they are not working in the fields). Control over food, land, and water is control, period. When control is consolidated, the money of the many—for we all need to eat—is funneled to the few, and the disparity continues to increase.

Observe the agricultural practices of a society and you will understand the foundational values and deeply rooted philosophies behind that society's dominant social and political systems. Our agriculture demonstrates the way we relate to the Earth, to the future, to fear, to money, and to each other. I say "our agriculture" because we all participate in agriculture every time we buy food. Ten years from now, will our agriculture reflect a culture of control and a stubborn belief in the binary of dominance? Or will our agriculture reflect a paradigm of partnership—a celebration of cooperation and respect for the inherent interconnectedness of everything?

Food is simple. Food is magic. Food is water, soil, seeds, animals, sunlight. When I eat my beef, it is a celebration. A celebration of the animal, first and foremost. A celebration of our sun, of clean water and pasture grass, of love and health. And, by the grace of this food, a celebration of my future. I know how lucky I am to raise so much of my own food and to buy much of what I don't raise from people and farms I know and trust, to be able to support these farmers directly, to know that their practices support the environment. I want our food systems to change, and change fast, because access to quality, ethically raised food shouldn't be determined by luck and circumstance. It shouldn't be a privilege. It is a fundamental human right.

An ethical food system is one that nourishes all of us and nurtures the planet, too. An ethical food system is a decentralized food system, in which money circulates through communities in a reciprocal relationship of mutual support instead of traveling up a ladder. An ethical food system ensures safe and supportive working and living conditions for those who perform farm work, and wages that honor their invaluable contribution to our lives. An ethical food system champions USDA-certified on-farm and mobile slaughter, so that animals are not unduly stressed by long-distance travel on the final day of their lives. An ethical food system grants autonomy and freedom of choice to farmers, ranchers, and consumers alike. An ethical food system is built through cooperation—and if ours is to evolve for the betterment of all life, we need the help of those outside the 1 percent of the population that works in agriculture for this transformation to take place.

have a slip of thin, blue paper—my carbon copy of the brand inspector's receipt—that I keep in an old cigar box and that shows I own Daisy. But the truth is that I belong to Daisy. Belonging—to a cow, to a place, to a person, to an idea—means you structure your life and consider your actions with their well-being in mind. Not just in mind, but prioritized—and it doesn't feel like sacrifice. I think back to the night I milked Daisy after her miscarriage—how my stress, my sleep deprivation, my shivering against the frigid air all disappeared while I milked Daisy in the dark, even though milking her was the last thing I would have chosen to do for kicks that night. Even though I gained nothing from it personally and couldn't even save the milk. I milked her because she needed me to milk her. And what I found, when I set to work, was that my devotion to Daisy trumped how cold I was—and more than that, it warmed me. My devotion to Daisy trumped how tired I was—and more than that, it energized me. My devotion to Daisy trumped how uncomfortable I was, huddled beneath her on the ice—but more than that, it gave me an almost superhuman endurance. Every devoted parent is familiar with this phenomenon. Every devoted partner and caretaker and friend knows it, too.

We all belong to the Earth. And we all belong to each other. For years, most of my account passwords have been variations on a quote by Mariah Makalapua: "May our actions reflect our priorities and not our fears." I've used this as a password because every time I type it, it's a reminder. And I need the daily reminder. There's no easy way I've found to live into this mantra, to reweave my conditioning, but I begin by asking, *"What can I give? What can we achieve together?"* When I ask these questions, the answers often come. And when they don't, I go outside and find Daisy. She is my answer. When I'm with her, time disappears, becomes irrelevant. One minute becomes an hour, an hour becomes one minute. I am unburdened. Am I presumptuous to believe the feeling is mutual? Sometimes, when I'm sitting in the grass, leaning against Daisy's shoulder in the afternoon sun, she swings her head around and rests it upon my chest, and falls asleep. It's difficult to breathe, pinned as I am between her rib cage and her fifty-pound head, but I don't mind. I sip shallow breaths as Daisy rests upon my heart, her eyelashes fluttering against my neck as she dreams.

ACKNOWLEDGMENTS

Immense gratitude to my agent, Stacey Glick, for believing in my work so completely, and my editor, Joanna Ng, for such precise and thoughtful guidance. My gratitude extends to the entire team at TarcherPerigee who helped bring this book into beautiful form.

Huge thanks to Karin Neff for the boost of confidence when I desperately needed it.

A deep bow of gratitude to Lynn George, Ervin Bader, and Lois Shirran—my life was changed for the better because our paths intersected.

My heartfelt thanks to everyone who has participated in the Food Bank Cooperation Donation* over the past several years, and all who have supported my work. I could not have learned and grown as I have without you.

Joslyn, much of this book grew from seeds you planted while you were here. Your extraordinary wisdom continues to guide me.

Mike, I cannot imagine what my life would be if I had never met you. I'm so glad I will never know. Thank you.

---

*Hello, you—yes, you, reading this—you can learn more about this cooperative effort, and join us if you like, here: https://bit.ly/cooperation_donation

REFERENCES

Memoirs are not generally footnoted, but I feel it is important to share my sources.

Much of the data shared in this book came directly from the United States Department of Agriculture: the USDA Economic Research Service and the 2017 USDA Census of Agriculture. The Census of Agriculture is conducted every five years; 2017 is the most recent at the time of this writing. I deeply appreciate the time that farmers and ranchers take from their busy lives to complete the census. This information is valuable to us all, and the 800 pages of compiled data is illuminating.

The 2017 USDA Census of Agriculture may be found here:

https://www.nass.usda.gov/Publications/AgCensus/2017/Full_Report/Volume
   _1,_Chapter_1_US/usv1.pdf

Following is a nonexhaustive list of sources and recommended reading for further exploration.

References

Alvarez, Ramon A.; Zavala-Araiza, Daniel; Lyon, David R. "Assessment of Methane Emissions from the U.S. Oil and Gas Supply Chain." *Science,* Vol. 361, Issue 6398, July 13, 2018.

Arsenault, Chris. "Only 60 Years of Farming Left If Soil Degradation Continues." Reuters, December 5, 2014.

Bailey, Robert. "Another Inconvenient Truth: How Biofuel Policies are Deepening Poverty and Accelerating Climate Change." Oxfam International, June 2008.

Baes, Christine. "Do Cows Produce More Methane Than Rotting Grass?" Quirks and Quarks, CBC Radio, March 1, 2019.

Brooks, Kathleen R; Raper, Kellie Curry; Ward, Clement E; Holland, Ben P; Krehbiel, Clint R. "Economic Effects of Bovine Respiratory Disease on Feedlot Cattle during Backgrounding and Finishing Phases." *The Professional Animal Scientist,* Vol. 27 Issue 3, 195–203, June 2011.

Campbell, John. "Liver Abscesses Still Significant Challenge for Cattle Industry." *The Western Producer,* January 8, 2015.

Capper, J.L. "Is the Grass Always Greener? Comparing the Environmental Impact of Conventional, Natural and Grass-Fed Beef Production Systems." Animals 2012, 2, 127–143.

Corah, Larry. "Cattle Sickness, Mortality Cost and the Search for Solutions." Angus Beef Bulletin, July 21, 2014.

Cummins, Ronnie. "Do You Know Where Your Meat Comes From?" Organic Consumers Association, May 24, 2018.

Dass, Pawlok; Houlton, Benjamin; Wang, Yingping; Warlind, David. "Grasslands May Be More Reliable Carbon Sinks Than Forests in California." Environmental Research Letters, Vol. 13 No. 7, IOP Publishing Ltd, July 10, 2018.

Donovan, Peter. "Measuring Soil Carbon Change: A Flexible, Practical, Local Method." Soil Carbon Coalition, October 2013.

Eisler, Riane. "The Chalice and the Blade: Our History, Our Future." HarperCollins, 1987.

FAO, IFAD, UNICEF, WFP and WHO. "The State of Food Security and Nutrition in the World 2019. Safeguarding Against Economic Slowdowns and Downturns." Rome, FAO, 2019.

Gauthier, Joki; Hoffman, Beth. "Behind the Brands: Food Justice and the 'Big 10' Food and Beverage Companies." Oxfam International, February 2013.

Harlan, Becky. "Digging Deep Reveals the Intricate World of Roots." *National Geographic,* October 15, 2015.

Howarth, Robert W. "Ideas and Perspectives: Is Shale Gas A Major Driver of Recent Increase in Global Atmospheric Methane?" *Biogeosciences,* 16, 3033–3046, 2019.

Hribar, Carrie. "Understanding Concentrated Animal Feeding Operations and Their Impact on Communities." National Association of Local Boards of Health, 2010.

Hristov, A.N. "Historic, pre-European Settlement, and Present-Day Contribution of Wild Ruminants to Enteric Methane Emissions in the United States." *Journal of Animal Science* 2012. 90:1371–1375.

Intergovernmental Panel on Climate Change [Masson-Delmotte, V., P. Zhai, H.-O. Pörtner, D. Roberts, J. Skea, P.R. Shukla, A. Pirani, W. Moufouma-Okia, C. Péan, R. Pidcock, S. Connors, J.B.R. Matthews, Y. Chen, X. Zhou, M.I. Gomis, E. Lonnoy, T. Maycock, M. Tignor, and T. Waterfield (eds.)]. "Global Warming of 1.5°C: IPCC Special Report." World Meteorological Organization, Geneva, Switzerland, October, 2018.

James, Ian; O'Dell, Rob. "Arizona's Next Water Crisis: Megafarms and Deeper Wells are Draining the Water Beneath Rural Arizona—Quietly, Irreversibly." *AZ Central,* December 27, 2019.

Johanns, Ann. "Iowa Cash Corn and Soybean Prices, 1925-2020." Ag Decision Maker, Iowa State University Extension and Outreach.

Johanns, Ann; Plastina, Alejandro. "Historical Costs of Crop Production, 1975–2020." Ag Decision Maker, Iowa State University Extension and Outreach.

Kerlin, Kat. "Grasslands More Reliable Carbon Sink Than Trees." Science & Climate, UC Davis, July 9, 2018.

Kessel, Jonah M.; Tabuchi, Hiroko. "It's a Vast, Invisible Climate Menace. We Made It Visible." *The New York Times,* December 12, 2019.

Kimmerer, Robin Wall. "Braiding Sweetgrass: Indigenous Wisdom, Scientific Knowledge and the Teachings of Plants." Milkweed Editions, October 2013.

Kittredge, Jack. "Soil Carbon Restoration: Can Biology do the Job?" Northeast Organic Farming Association, 2015.

References

Larson, C.L.; Reed, S.E.; Merenlender, A.M.; Crooks, K.R. "Effects of Recreation on Animals Revealed as Widespread through a Global Systematic Review." PLoS ONE 2016, 11(12): e0167259.

Leahy, Stephen. "Insect 'Apocalypse' in U.S. Driven by 50x Increase in Toxic Pesticides." *National Geographic,* August 6, 2019.

Leonard, Christopher. "The Meat Racket: The Secret Takeover of America's Food Business." Simon & Schuster, January 2014.

Mann, Charles C. "1491." *The Atlantic,* March 2002.

Mann, Paul. "Can Bringing Back Mammoths Help Stop Climate Change?" *Smithsonian Magazine,* May 14, 2018.

Market Wrap, ISec Report. "Global Agrochemical Industry: Muted Prices Drive Consolidation." ICICI Securities, September 29, 2017.

McCluskey, Jill; Wahl, Thomas; Quan, Li; Wandschneider, Philip R. "U.S. Grass-Fed Beef: Marketing Health Benefits." *Journal of Food Distribution Research* 36(3), November 2005.

Montgomery, David R. "Dirt: The Erosion of Civilizations." University of California Press, May 2007.

Motamed, Mesbah. "Federal Commodity Programs Price Loss Coverage and Agricultural Risk Coverage Address Price and Yield Risks Faced by Producers." United States Department of Agriculture, Economic Research Service, August 6, 2018.

National Farmers Union, Farm Crisis Center: www.farmcrisis.nfu.org

National Oceanic and Atmospheric Administration, Earth System Reasearch Laboratory. "Trends in Atmospheric Methane."

National Parks Service. "Tallgrass Prairie: A Complex Prairie Ecosystem." 2018.

Niman, Nicolette Hahn. "Defending Beef: The Case for Sustainable Meat Production." Chelsea Green Publishing, October 2014.

Nosowitz, Dan. "Landmark 20-Year Study Finds Pesticides Linked to Depression In Farmers." Modern Farmer, November 7, 2014.

Ontl, T.A.; Schulte, L.A. "Soil Carbon Storage." *Nature Education Knowledge,* 3(10):35, 2012.

Oswald, Richard. "Letter from Langdon: Cheaper to Buy Than to Grow." *The Daily Yonder,* January 3, 2017.

References

Penniman, Leah. "Farming While Black: Soul Fire Farm's Practical Guide to Liberation on the Land." Chelsea Green Publishing, October 2018.

Pittman, Craig. "The Clock is Ticking on Florida's Mountains of Hazardous Phosphate Waste." *Sarasota Magazine,* May 2017.

Pollan, Michael. "Deep Agriculture." Long Now Foundation, May 5, 2009.

Ranchers-Cattlemen Action Legal Fund United Stockgrowers of America Class Action Complaint, Case No. 19-1222. Filed May 7, 2019; United States District Court, District of Minnesota. https://www.r-calfusa.com/wp-content/uploads/2019/05/Cattle-complaint.pdf

Raygorodetsky, Gleb. "Indigenous Peoples Defend Earth's Biodiversity." *National Geographic,* November 16, 2018.

Regrarians Media. "Polyfaces, A World of Many Choices." 2015.

Ringgenberg, Wendy Jeannette Wehrman. "Trends and Characteristics of Occupational Suicide and Homicide in Farmers and Agriculture Workers, 1992-2010." Master of Science Thesis, University of Iowa, 2014.

Rodale Institute. "Regenerative Organic Agriculture and Climate Change, A Down-to-Earth Solution to Global Warming." 2019. https://rodaleinstitute.org/wp-content/uploads/Regenerative-Organic-Agriculture-White-Paper-RodaleInstitute.pdf.

Savory, Allan. "Holistic Management." Island Press, 2006.

Sobrevila, Claudia. "The Role of Indigenous Peoples in Biodiversity Conservation." World Bank, May 2008.

Thorbecke, Mariko; Dettling, Jon. "Carbon Footprint Evaluation of Regenerative Grazing at White Oak Pastures." Quantis, February 25, 2019.

Turkewitz, Julie. "Who Gets To Own The West?" *The New York Times,* June 22, 2019.

United States Department of Agriculture, Agricultural Marketing Service. "Country Of Origin Labeling (COOL)." (ams.usda.gov)

United States Department of Agriculture, Economic Research Service. "Feedgrains Database." "Corn and Other Feedgrains." "Soybeans and Oil Crops." "Food Markets and Prices." "Food Prices and Spending." (ers.usda.gov)

United States Department of Agriculture; Economics, Statistics and Market Information System. "Cattle on Feed." (Reports dated March 8, 2019 through January 24, 2020).

United States Department of Agriculture, Food Safety and Inspection Service. "Meat and Poultry Labeling Terms." 2015. (fsis.usda.gov)

United States Department of Agriculture, National Agricultural Statistics Service. "2017 Census of Agriculture, Full Report." April 2019. (nass.usda.gov)

United States Environmental Protection Agency. "Inventory of U.S. Greenhouse Gas Emissions and Sinks: 1990–2018; Chapter 5: Agriculture;" "Sources of Greenhouse Gas Emissions;" "Overview of Greenhouse Gasses: Methane Emissions." (epa.gov)

Wallace-Wells, David. "The Uninhabitable Earth: Life After Warming." Tim Duggan Books, February 2019.

Wise, Timothy A. "Identifying the Real Winners from US Agricultural Policies." Global Development and Environment Institute, Tufts University, December 2005.

Zielinski, Sarah. "Earth's Soil Is Getting Too Salty for Crops to Grow." *Smithsonian Magazine,* October 28, 2014.

Lastly, if you or someone you know has been poisoned, call Poison Control (1-800-222-1222) and 911. Do not administer or ingest charcoal indiscriminately as it can cause additional harm, and do not ingest charcoal from a fireplace or barbecue!

ABOUT THE AUTHOR

Shreve Stockton is an award-winning photographer and author of *Eating Gluten Free*; *The Daily Coyote*, a memoir of raising an orphaned coyote; and *The Daily Coyote: Ten Years in Photographs*. She is the founder of Star Brand Beef, devoted to prioritizing the humane treatment of animals and ethical stewardship of the land. She lives in Wyoming with her Farmily—cows, bulls, cats, dogs, horses, honeybees, chickens, a coyote, and a cowboy.

She can be found online at shrevestockton.com